100 Ideas for Secondary Teachers:

Supporting Students with Numeracy Difficulties

Patricia Babtie and Sue Dillon

BLOOMSBURY EDUCATION

LONDON OXFORD NEW YORK NEW DELHI SYDNEY

BLOOMSBURY EDUCATION
Bloomsbury Publishing Plc
50 Bedford Square, London, WC1B 3DP, UK

BLOOMSBURY, BLOOMSBURY EDUCATION and
the Diana logo are trademarks of Bloomsbury Publishing Plc

First published in Great Britain, 2019

A catalogue record for this book is available from the British Library

ISBN: PB: 978-1-4729-6109-9; ePDF: 978-1-4729-6108-2;
ePub: 978-1-4729-6107-5

2 4 6 8 10 9 7 5 3 1 (paperback)

Typeset by Newgen KnowledgeWorks Pvt. Ltd., Chennai, India
Printed and bound in Great Britain by CPI Group (UK) Ltd, Croydon CR0 4YY

To find out more about our authors and books visit
www.bloomsbury.com and sign up for our newsletters

Contents

Acknowledgements   vi
Introduction   vii
How to use this book   viii

Part 1: Study skills and exam technique   **1**
1. GET SET   2
2. Five-minute magic   3
3. Daily data   4
4. What to do with all that paper   5
5. Exam tamer   6
6. Keep breathing   8

Part 2: Number basics   **9**
7. Numeracy basics   10
8. Real numbers   11
9. The estimating game   12
10. The balancing act   14
11. Fact teams   16
12. Signs and symbols   18
13. Rounding numbers   20
14. Rounding practice   22
15. Prime time   24
16. Hunt multiple factors   26
17. 60-second prime factors   28
18. Square numbers   30
19. Cryptography – crack the code   31

Part 3: Fractions, decimals and percentages   **33**
20. Fraction foundations   34
21. Fraction bars and lines   35
22. Fractions: the area model   36
23. Numerator and denominator effects   38
24. Exciting equivalence   40
25. Fraction comparison calculations   42
26. Simplifying fraction models   44
27. Fraction wall   46

28	Fractions of sets	48
29	Percentages	50
30	Decimals	52
31	Same fraction, different form	54
32	Four fractions to make a whole	56
33	Memorable fraction forms	58
34	Improper and mixed fractions	59
35	Adding and subtracting fractions	60
36	Multiplying by fractions	62
37	Dividing by fractions	64

Part 4: Approximation, rounding and significant figures — **67**

38	The rounding decider	68
39	The limits of rounding	70
40	Significant numbers	72
41	The Seven Summits	74
42	The significance of significant figures	76

Part 5: Roots, powers and standard form — **77**

43	Rooting out square roots	78
44	More rooting around	80
45	Positive and negative square roots	82
46	Power in the chain	84
47	Multiplying and dividing powers	85
48	The power of geometric growth	86
49	Introducing standard form	88
50	Multiplying and dividing with standard form	90

Part 6: Ratio and proportion — **91**

51	Ratio and proportion	92
52	Cooking it up	94
53	Plan to scale it down	95
54	Map reading	96
55	Human scale	97
56	Universal conversion grid	98
57	Rates of change: percentage increase	100
58	The rule of 72	101
59	The sale price: percentage decrease	102
60	Money surprises	104

Part 7: Algebra — **107**

61	Why collect terms?	108
62	Algebra: substitution	109
63	Expanding brackets	110

64	Algebra: the beginnings	112
65	Sign aware	114
66	Variable, coefficient, operation, constant	116
67	What 'like' is like?	117
68	The short hand of algebra	118
69	So why do we use these letters?	119
70	Playing with multiplication	120
71	What to do with exponents?	121
72	Baffling brackets	122
73	Bracket practice	123
74	Contraction and expansion	124
75	Express and equate	126
76	Construct an equation	127
77	Which operation?	128
78	Equations again	129
79	Algebra is a balancing act	130
80	Two sides with an unknown	131
81	The sequence of working	132

Part 8: Measurement basics | | **133**
82	Memorable measures	134
83	How big is that number?	135
84	In equal measure	136
85	Average	138
86	Weighty matters	140
87	Graph stories	141
88	Time and motion	142
89	The language of clocks	144
90	Digital and analogue time	146
91	The calculator trap	148
92	Talking time	150
93	Distance-time graphs	152
94	Temperature	154

Part 9: Interesting investigations | | **157**
95	Squares and odd numbers	158
96	Number triangles	160
97	From one to eternity	162
98	Into another dimension	164
99	Investigating pi	165
100	Growth drama	166

Acknowledgements

The joy of teaching students who find numbers difficult is sharing their relief and excitement when they discover they can master them. Fear prevents many students from learning maths so dispelling anxiety is a crucial part of the ideas in this book. Thank you to all the students who have worked with us over the years to enable us to figure out what works for them.

Multi-sensory learning is key to unlocking the magical world of numbers and we have been lucky to work with some inspirational people developing this approach. Prof Brian Butterworth unlocked the neurological basis of dyscalculia through research in Cognitive Neuroscience at UCL. At Emerson House, the late Dorian Yeo worked with him to translate those findings into effective teaching methods and develop them with other educators, including the authors. Patricia and Jane Emerson wrote books in order to make the ideas as widely available as possible.

There are so many people who have shared ideas and encouraged us that it is not possible to thank everyone individually. Patricia hugely appreciates the collaboration of colleagues in a variety of schools as well as the discussions and contributions of those who attended lectures and seminars. Sue is grateful to colleagues at Thomas's Battersea for all their support and encouragement over many years.

A huge thank you to Hannah Marston, our superbly patient editor at Bloomsbury, whose incisive editing has clarified our writing. And thank you to those who have read and commented on various drafts: John Babtie, Lucinda Barry and Ann Fitzjohn-Sykes.

For those interested in finding out more about the links between neuroscience and education, visit the website for the Centre for Educational Neuroscience at www.educationalneuroscience.org.uk.

Introduction

Learning maths is about finding things out, fostering an enquiring mind and developing the imagination. 'Tell me what you see' are the most powerful words in teaching maths. They encourage students to look at the world, analyse information and discuss it. They 'tell', that is communicate, their ideas using models, diagrams, charts and graphs as well as words and numbers. Mistakes present opportunities to revise and refine thinking. It is best to work in small groups so that each student has a chance to talk about their ideas.

Start from first principles and show students how to uncover the concepts behind the numbers. Do check that they know what basic counting means. Too often this is taken for granted yet often numeracy difficulties start here (see Part 2).

The numerical skills needed for maths are also essential for the sciences, geography and economics. It is less obvious that mathematical understanding also helps in art and history by developing skills such as observation, analysis, logic, and the ideas of scale and magnitude.

All the sciences require mathematical proficiency in every area covered in this book. Bring the numbers to life through examples from the real world, as well as modelling wherever possible. Measurements of all kinds make more sense when related to the human experience. Geographers will find applications of scale and ratio in drawing plans and reading maps. In history lessons, consider the evolution of ideas such as telling the time, or the influence of pi.

The emphasis on exam results sometimes overshadows the fact that numeracy is also an essential life skill. Numerical data drives many important life choices. Unfortunately, data can be represented in mischievous ways to obscure the true meaning, so it is essential that students learn to assess data critically.

Students who struggle with numeracy in secondary schools are often afraid, so it is vital that they enjoy the discovery process rather than competing to get answers as quickly as possible. Calculation is important but what is more important is that students learn to be flexible thinkers who can use numbers to solve problems in a variety of situations.

The key to success lies in developing a positive attitude where students model, draw, discuss and write down ideas without fear. Have fun!

How to use this book

This book includes quick, easy and practical ideas for you to dip in and out of, to help you in supporting students with numeracy difficulties.

Each idea includes:

- a catchy title, easy to refer to and share with your colleagues
- an interesting quote linked to the idea
- a summary of the idea in bold, making it easy to flick through the book and identify an idea you want to use at a glance
- a step-by-step guide to implementing the idea.

Each idea also includes one or more of the following:

Teaching tip	Taking it further	Bonus idea ★
Practical tips and advice for how and how not to run the activity or put the idea into practice.	Ideas and advice for how to extend the idea or develop it further.	There are 26 bonus ideas in this book that are extra-exciting, extra-original and extra-interesting.

Share how you use these ideas and find out what other practitioners have done using **#100ideas**.

Online resources also accompany this book. When a resource is referenced, visit www.bloomsbury.com/100-ideas-secondary-numeracy-difficulties to find extra resources, catalogued under the relevant idea number. Here you can also find the full list of website addresses mentioned in the book.

Study skills and exam technique

Part 1

GET SET

'Develop a healthy work–life balance to set a pattern for a happy life.'

Help students to gain control of their lives so that they have enough time to work, exercise, socialise and sleep.

Teaching tip

Discuss with students the following points:

- Do you take responsibility for your own learning?
- How many hours are in a week?
- Which key things do you want and need to do each week? This includes eating meals, exercising and sleeping.
- How much time is then available for studying?

There are detailed suggestions for learning about time management in Idea 3 and effective organisation in Idea 4.

Taking it further

The average teenager in the UK spends nearly two hours a day on social media. Together with the students, work out how much of the school year they spend on social media.

Students with numeracy difficulties often have underlying organisational problems. They find it hard to analyse information in order to solve problems. Learning that structure makes for a happier life is a powerful motivator to encourage planning and organisation in all academic areas. This activity is also a good starting point for making it clear that it is important to learn to tell the time, something that some students find very difficult.

Time spent improving study habits is really worth it. Students will have heard it all before but now you need to make sure that they act on it. The mnemonic 'GET SET' underlines the fact that this is about preparation for studying. It is also preparation for life.

Effective learning starts with effective living: health, time to work and play, curiosity and engagement. Teach students the following acronym, making sure they have a clear understanding of each aspect:

G – goal: short, medium or long term
E – energy: healthy diet, enough exercise, enough sleep
T – time: time to think, absorb and live
S – structure: schedule, filing
E – effective effort: being organised to start studying, motivation, filing notes
T – targets: realistic, measurable targets to achieve each goal

Five-minute magic

'The first step in effective study is to know what you need to learn – in broad outline.'

What do you need to study? Research shows that students often don't know what topics are included in the syllabus. Mathematical concepts build on prior information so it is vital that students with numeracy difficulties are clear about what they need to learn.

Knowing what you know, and what you don't know, is crucial to effective study – whatever the subject. Spring this exercise on students with no warning. There is a huge difference between recognising and remembering. Make sure that you do not use the word 'list' in your instructions; students need to record the information in the way they choose, whether written or pictorially.

Students summarise the courses they will be studying for the year. Write the name of each subject and the key topics for each subject. Give students five minutes to complete the task.

Next, students discuss their findings:

- Did the students find the task easy?
- Do they think that they recalled all the key points?
- How did they organise the information?
- What are the advantages of a list or a mind map?
- Did they use diagrams as well as words?

Provide a summary of the maths syllabus for the year in both a linear and spatial format. Ask students to find what topics they omitted from their own summary.

Teaching tip

Introduce a 60-second habit at the start and end of each lesson. Give students the lesson topic and ask them to write two points they know about it. If they struggle to remember, then they may not have fully understood the basics. At the end of the lesson, give them one minute to write down three key points.

Taking it further

Discuss the difference between recall and recognition. Too often students see information and say, 'I know this' when they are actually only recognising that they have seen it before. Recall is the ability to retrieve information from memory when required. Too often it is associated with rote learning, whereas learning by understanding the concept and its usefulness is much more effective. Useful memories involve linking new information to what is already stored in memory.

Daily data

'Plan your time. Stick to your plan. Make it a habit.'

Time management is key to a successful, happy life. Students need to plan what they are going to study each day and how long it will take. They underestimate the amount of time they spend on electronic devices each day. Are they pawns of the big tech companies? Or masters of their electronic devices?

Habit audit

First, students establish what their work habits are by keeping a record of exactly how they spend their time every day.

They make a timetable – on paper, not a computer – and mark half-hourly slots. They record what they are doing every 15 minutes. They write the time they start an activity but make sure they are sensible about it. If they are doing a sustained activity, e.g. a football game, they only note the start and end time. It is crucial they also note any short changes, including every 'quick check' of smartphone alerts as well as time spent on social media.

At the end of the day, the students work out how long they have spent on lessons, checking their phone, online, eating, taking exercise and doing homework or other independent study. What can they cut down to give themselves one hour extra a day?

Plan your study

Now students use the following process to plan their study time:

- What do you *want* to do each day?
- What do you *have* to do each day?
- How long will it take you every day?
- Draw up a timetable for each day.
- Draw up a timetable for the week.

What to do with all that paper

'"Where is your homework?" asks the teacher. "At the bottom of my bag," replies the student.'

Organising filing helps organise thoughts. Many students with organisation and memory difficulties have little idea of how to collate and keep their work in an organised way so that they can easily access it for review and revision.

A good filing system will help memory. The brain links new material to information that is already in memory. Apply the same principle to filing by encouraging students to have one folder for each subject and clearly identified subtopics within the subject folder.

It is helpful for students to create a colour-coded sequential filing system, which will increase students' recall and understanding as to where they are in their learning.

- Show students a visual map of topic headings that they will study. For example:

Teaching tip

Develop the habit of filing paper immediately. End each lesson five minutes early and insist that students file the information from the lesson in a day file (see 'Taking it further') so it is ready for review before being transferred to the subject file. Start the next lesson by asking students to confirm that they have done the appropriate filing.

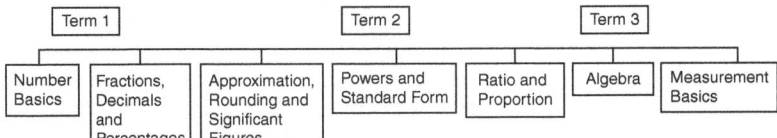

- Discuss each subtopic with the students.
- Colour code the topic headings on the timeline. Use a colour for each topic and its subtopics.
- Suggest that students use a lever-arch file.
- Each student divides their file by topic with file dividers that match the topic colours.
- Ask students to use bold, large headings and relate pages to one another.
- Students add an introduction to each topic listing subtopics and page numbers.

Taking it further

As the lever-arch file becomes fuller and heavier, students may wish to use a thinner day file to take home pages of work.

Exam tamer

'Control your time; don't let time control you.'

Too often teachers, tutors or parents organise teenagers' lives for them. Young people need to take control of and responsibility for their own learning. In this activity, students discover how many marks they can achieve in ten minutes, then they practise daily to improve their score in a calm and measured way.

Teaching tip

The important message is that this technique only works if the student takes charge. They need to practise during independent study. Doing it in school at a time designated by the teacher removes the responsibility from the student.

Students often feel they are working against the clock, rather than with the clock. They need to learn how to work at a steady rate and gradually become more efficient. Teach them to use a timer to develop an internalised sense of time and chart progress on test questions. It requires effort on the part of the student but the reward is better exam results, and much more importantly, it sets a pattern for life.

The teacher introduces the idea of measured practice to the whole class. Thereafter, students take responsibility for their own learning: they practise every day and keep a record of what they achieve. The aim is to make exams less stressful. Students should start by attempting to answer questions worth five marks in ten minutes. They gradually learn to work more efficiently. The goal is to achieve one mark per minute on an exam paper.

First, as a whole class, ask how many marks students can achieve in ten minutes:

- Provide a question paper with questions worth a total of ten marks. This allows more able students to keep engaged in the task.
- Explain that the aim is for each student to establish their own benchmark.
- Set a timer for ten minutes.
- Ask students to stop when the timer goes off and record how many marks they attempted.
- Mark the work and give feedback in private.

Now encourage the students to practise independently by following these instructions:

- Practise maths questions for ten minutes a day.
- Draw up a table to record your progress.
- At the start of each week, plan when you will do the ten minutes of practice.
- Set your own realistic target for improvement each week, e.g. improve by one or two marks each week.
- Check you have what you need before you start.
- Set the timer for ten minutes.
- Work steadily.
- Stop when the timer goes off.
- Record how many marks you've attempted.
- Later record how many answers are correct.

It is essential to use a timer, not a smartphone or a clock. The experience of working uninterrupted, until the timer sounds, encourages focus and develops a sense of what is achievable in the time. Checking a watch to see how much time is left breaks the train of thought.

Taking it further

Knowing what a certain length of time 'feels' like is invaluable in exams. It helps students pace themselves and reduces stress. However, it is an important life skill. Mark McCormack, the hugely successful sports and celebrity management agent, said that time management was the key to his success.

Keep breathing

'This amazing technique helped me to calm down and stop panicking during exams: remember to breathe.'

Develop the habit of relaxed breathing to combat stress and improve exam results. Students who practise this simple routine for at least five minutes a day, every day, find it reduces anxiety and helps them focus.

Teaching tip

Rather than talk students through the session, use recorded instructions. Download the free guided meditation 'Breathing Meditation (5 Minutes)' from the University of California Los Angeles (UCLA) website: www.uclahealth.org/marc/mindful-meditations.

Taking it further

Impress on students that they need to practise daily to establish the habit of calm breathing. Only when it is a habit will they find it easy to implement in stressful situations. Use the time whilst exam papers are handed out to practise calm breathing; it helps focus attention and improves confidence.

We should all remember to breathe properly. It sounds obvious; after all, we breathe without thinking. However, anxiety and stress cause people to take shallow breaths or to hold their breath without even realising it. This in turn increases the anxiety that affects clear thinking. Erratic breathing is a common trait amongst anxious examinees.

The following breathing exercise can be done as a whole-class activity or individually. It is imperative that breathing sessions are non-judgemental. Some students will find it easier than others; do not compare them. Guide students through a five-minute session of breathing to encourage calm, relaxed, deep breaths. Aim to observe the in-breath and out-breath without trying to change it. Talk calmly and slowly.

'I am going to set a timer for five minutes. Sit comfortably with your back supported by the chair and arms loosely folded on your lap. Close your eyes. Focus your attention on your breathing. Feel the air come in through your nose. Relax as the air flows into your lungs. Feel your abdomen rise as the air expands the lungs. Then, feel the air flow out of the lungs and out through your mouth. Don't force your breathing. Simply be aware of the process of breathing.'

Number basics

Part 2

Numeracy basics

'You cannot build further knowledge without basic number ability.'

Students who have weak working memory difficulties and perhaps have a history of struggling to understand maths concepts or to learn maths facts will need to review the basics before they can access higher maths concepts such as algebra.

The ideas in this book assume that students understand the following foundations of numeracy:

- counting (cardinal and ordinal value of numbers)
- base ten structure of the number system
- the principle of exchange
- place value
- a sense of the size of a number
- approximation and estimation
- key facts: addition and subtraction
- multiplication and division by 10 and by 100 and by 1,000
- partitioning a number into components
- concept of multiplication as repeated addition and as area
- concept of division as grouping or sharing
- division with remainders
- the difference between discrete (counting) numbers and continuous (measuring) numbers
- the relationship between a number track and a number line
- the ability to use a ruler to measure correctly and its relationship to a number line.

This section of the book develops these basic concepts further to ready students for numeracy in the secondary curriculum.

Real numbers

'Our ancient ancestors invented numbers because they needed them to count and measure things.'

Many students struggle with calculations because they do not understand the conceptual difference between discrete (counting) and continuous (measuring) numbers.

Natural numbers: These are the positive whole numbers used for counting objects.

Integers: These are all the whole numbers including zero and negative numbers.

Rational numbers: These are the numbers used for measurement. They include whole numbers and all the values between them. Any rational number can be written as a fraction in which the numerator and the denominator are whole numbers. They include integers.

Number line: All rational numbers can be shown at a precise point on a number line.

Irrational numbers: These numbers cannot be shown as a ratio between two whole numbers. An important irrational number is π (pi) (see Idea 99).

Real numbers: All the rational and irrational numbers are called real numbers.

Imaginary numbers: If an imaginary number is squared, the result is a negative number. Imaginary numbers have practical importance in areas such as electricity supply.

Taking it further

Explore the history of the invention of numbers and depict them on a timeline along a classroom wall.

The estimating game

'Probably the most important maths question is: does the number make sense? Having a sense of the size of a quantity helps.'

The ability to quickly size up a number or make quick calculations is a fundamental skill. It is called estimation. Some think it old fashioned but it is invaluable for checking that calculations make sense. This activity develops confidence in estimation and provides an opportunity to visualise numbers in a linear and a spatial format, which is essential for calculations involving area.

The estimating game develops confidence in judging the size of quantities. All objects in one game need to be the same size and scale.

The basic linear game

- Students play in small groups.
- Each group has a container with more than 100 objects.
- Each player draws up a score sheet to record the estimates of all the players.
- Players take turns to scatter the objects. First, Player A scatters a quantity of objects on the table. They allow a few seconds for players to look at the objects, then cover them with a sheet of paper.
- Players say and record their estimates.
- Player A counts the objects into a line, leaving a small gap between each group of ten.
- The winner is the player whose estimate is closest to the actual number of objects.

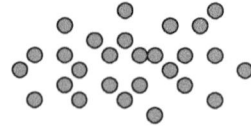

	Estimate of number				Actual	
	Player A	Player B	Player C	Player D	Number	Winner
Game 1	23	31	22	29	27	Player D

The spatial model

Once students are reasonably accurate at estimating quantities and are confident with linear representation of numbers, introduce the idea of quantities in a spatial array. The spatial arrangement paves the way for an understanding of area and also provides a solid foundation on which to build knowledge of percentages (see Idea 29).

- Start with a container of about 60 counters. Do not count them in front of the students.
- As in the basic game, students take turns to scatter a handful of objects, observe them briefly and then each record their own estimate of the number of objects.
- This time, the player who scattered the objects arranges them in an array with ten counters in each row.
- The winner is the person whose estimate is closest to the actual number of objects.

Bonus idea ★

Number judgement

Students are given a number and try to take the correct quantity of counters out of a tub. Each player has five turns and records their error score for each turn.

- Player A rolls two 0–9 dice, makes the larger target number out of the two digits and records it.
- Allow the player five seconds to take a quantity of counters out of the tub.
- Player B counts the objects as above and calculates the difference between the target number and the actual number.
- The winner is the player with the lowest error score after five rounds.

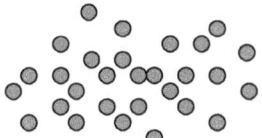 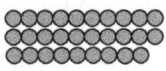

	Estimate of number				Actual	
	Player A	Player B	Player C	Player D	Number	Winner
Game 1	23	31	28	32	29	Player C

The balancing act

'Oh! Now I can see that the equal sign shows the relationship between two mathematical ideas. I thought that the equal sign meant "write the answer".'

The equal sign states that the quantity on one side of the = has exactly the same value as the quantity on the other side. Because value is an abstract quality, students may have difficulty grasping this concept. Use a balance for a striking demonstration of the concept. Then, students explore the consequences of changing the amount on one side and what action they need to take to compensate to keep the balance.

Teaching tip

If you do not have balance scales, hang a wire coat hanger on a door knob to make a simple balance. You will need clear plastic bags to hold the objects.

Use a simple balance scale, objects that all have a uniform weight and a paper tissue. Place seven objects on the left-hand side of the balance. Conceal three further objects in the tissue and write the letter A on the tissue. Put the tissue-covered objects on the right-hand side of the balance. Discuss what you need to do to make the scales balance.

When it is agreed that you need to add some more objects to the right-hand side, students draw a diagram and write an equation to pose the question.

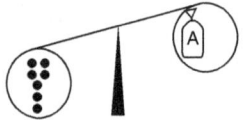

$$7 = A + ?$$

Pose the question: what do you need to add to make the scale balance? Add objects, one at a time, to the right-hand side until the scales balance. Write the answer to the question:

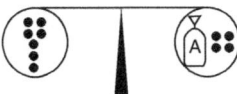

$$7 = A + 4$$

Now ask: how many objects are under the tissue? Discuss what action to take. Now you know that *A* plus 4 makes 7, you can subtract 4 objects from each side. The amount that remains will have the same value as *A*. Write the equation to show the process:

$$7 - 4 = A + 4 - 4$$
$$3 = A$$

The result is $A = 3$. Reveal the concealed quantity to check the result.

Students then work in pairs to do the same activity with larger numbers. Once students can confidently explain what they are doing and why, replace the letter *A* with the symbol *x* for unknown.

Taking it further

Use a question mark (?) to denote the unknown. Students read it as 'what', which draws attention to what they need to find. This helps them understand algebra where ? can be replaced by *x*, which is the conventional symbol for an unknown in algebra.

Fact teams

'There are lots of ways of describing one arithmetical relationship. And it works for algebra too.'

Help students to discover that the same arithmetical relationship can be described in several different ways.

In this activity, students model a number, split it into two components and draw a triad diagram, then write all the equations that the triad represents. A triad shows the relationship between a number and its components by using an inverted V shape to link the number to its parts.

Make it clear that addition and subtraction are inverse operations. The result of adding two numbers can be reversed by subtracting one of the numbers from the total. The order in which quantities are added does not affect the result; changing the order in subtraction gives a different result.

Investigation

- Students work in pairs and take turns to model and explain what they are doing. They start with small numbers, then practise with larger numbers.
- They use counters to split 5 into 2 and 3. They then draw the triad.

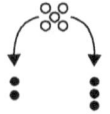

Finally, they write all the equations that describe the relationship between 5, 2 and 3:

$5 = 2 + 3$ $\quad$ $2 + 3 = 5$ $\quad$ $5 - 2 = 3$ $\quad$ $3 = 5 - 2$

$5 = 3 + 2$ $\quad$ $3 + 2 = 5$ $\quad$ $5 - 3 = 2$ $\quad$ $2 = 5 - 3$

The generalised equation

Now students generalise the concept by using letters to represent any number. First, they draw the triad:

Then, they write all the equations that describe the relationship between a, b and c as shown in the triad:

$a = b + c$ $b + c = a$ $a - b = c$ $c = a - b$

$a = c + b$ $c + b = a$ $a - c = b$ $b = a - c$

Taking it further

Understanding the concept of partitioning helps students develop flexible approaches to calculations with larger numbers. It is also essential for calculating effectively with number lines (see online resources).

Signs and symbols

'Misunderstanding and muddling up the equal sign and inequalities signs were common problems until I did this demonstration.'

The equal sign and the inequalities signs are a quick way of showing the relationship between quantities or mathematical expressions. However, some students are confused about what they mean.

Taking it further

Use the inequalities signs to compare three values that are unequal. This has an important role in defining the bounds, or limits, of numbers that have been rounded (see Idea 39).

Investigate the structure of the equal sign and other signs associated with it by modelling them with rulers.

Equal (=)

- Hold two identical rulers horizontally and parallel with a distance of about 5 cm between them to model the equal sign.
- Ask students to use their own words to describe what they see.
- Summarise contributions and write them on the board. Key elements are: two sticks, exactly the same length, horizontal, parallel, exactly the same distance apart.
- Discuss why two short parallel lines are a good choice for an equal sign to compare two things of the same value.
- Students model examples of situations that compare two things of the same value though the form may be different, and draw diagrams to include the equal sign. E.g.:

Inequalities (<>)

- Model the equal sign then move the rulers so that two ends come together and the other two move wider apart.

- Discuss the model as above. Now students need to describe how the gap on one end is smaller than the gap on the other.
- Encourage varied comparative language: unequal, not the same, more than, less than, greater than, smaller than, bigger than and fewer than. Write contributions on the board and discuss them.
- Students then compare values that are unequal using models. E.g.

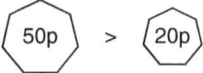

Other comparative signs

Students work in pairs to model and draw other signs showing relationships between numbers, including those in the table below.

They discuss why each sign is an effective summary of the meaning.

Sign	Meaning
≠	not equal
≤	equal to or less than
≥	equal to or greater than
≡	congruent, exactly equal
≈	approximate

Rounding numbers

'Rounding numbers becomes second nature on the number line.'

Sketching numbers on a number line helps students to grasp the significance of rounding and makes it clear when a number is rounded up or down.

Rounding a number simplifies it so it is easier to understand and to use. The number line image makes it clear why numbers round up from the midpoint and round down below the midpoint.

Sketch a number line from 0 to 10, mark the midpoint and then mark the whole numbers. Use an arc to show the distance from 5 to 10 above the line. Discuss the rule that 5 and above round up and numbers less than 5 round down.

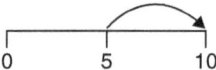

Rounding to the nearest 10

Ask students to:

- Sketch a number line from 0 to 100 and mark the decade numbers.
- Apply the rule to round the following numbers to the nearest ten: 23, 35, 57, 84.
- Find each number on the number line. It helps to mark the relevant midpoints.
- Write numbers that round down below the line and numbers that round up above it.
- Write a statement for each number.

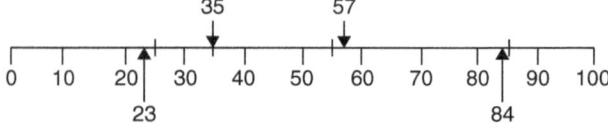

23 rounds down to 20.
35 rounds up to 40.

57 rounds up to 60.
84 rounds down to 80.

Rounding to the nearest 100

In the task above, the decider was the last digit in the number but it may not always be so. The decider is the digit that dictates whether you round the number up or down (see Idea 38). Students need to analyse the information and identify the appropriate decider. To model this:

Use a number line from 0 to 1,000 to round the following numbers to the nearest 100: 236, 467, 650, 748.

Discuss which digit in the number is the decider and why. The position of the number on the number line makes it clear that the tens digit is the decider.

Taking it further

Instil the habit of working out rough estimates before doing calculations and show these on empty number lines.

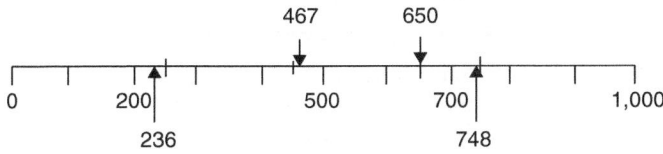

Rounding practice

'Add a competitive element to encourage rounding practice.'

Play the rounding game to practise rounding numbers to the nearest ten, or power of ten.

The rounding game hones the skill of quickly judging the relative size of numbers and placing them on a number line. It is important that students draw their own number lines rather than measuring them out.

Rounding to the nearest 10

- Each player sketches a number line and marks the decade numbers to 100.
- Players take turns to roll two 0–9 dice.
- On each turn, the player makes a two-digit number from the numbers they have rolled and rounds it to the nearest ten. Then, they consider the second possibility. They record both numbers.
- The player chooses one of the rounded numbers and crosses it out on the number line.
- The winner is the first person to cross out three consecutive decade numbers.

73 rounds to 70.
37 rounds to 40.

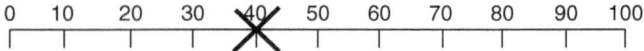

Rounding to the nearest 100

- Each player sketches a number line and marks the hundreds numbers to 1,000.
- Players take turns to roll three 0–9 dice.
- On each turn, the player makes all possible three-digit numbers from the numbers they have rolled and proceeds as above. Note that now there will be six possible numbers to choose from.

Taking it further

Play the same game on an empty number line. Only the beginning and end points are marked. This develops the ability to judge the relative value of quantities.

The player chooses 725, which rounds to 700.

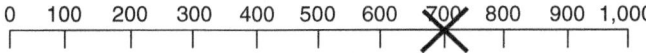

0   100   200   300   400   500   600   700   800   900   1,000

Prime time

'A great way to identify the primes to 100 as well as practising multiples of numbers.'

A prime number has two factors — itself and one. Investigate this idea with students on a 100 square.

Taking it further

Discuss why 2 is the only even prime number. This is an opportunity to revise the concept and definition of odd and even numbers.

In the third century CE, a Greek mathematician devised a way of finding prime numbers. It is called the Eratosthenes Sieve in his honour. Students can use this method to identify all the prime numbers to 100.

Students work individually to identify all the prime numbers on a 100 square. They should put a red circle around a prime number and use a blue pencil to cross out numbers that are not prime. To help them, provide the following instructions:

- Start by crossing out number 1 as it is a factor of all the numbers but it is not a prime number as it has only one factor: itself.
- Circle number 2 in red.
- Work systematically and cross out each number that is a multiple of 2.
- Circle the number 3 in red.
- Work systematically and cross out each number that is a multiple of 3.
- Find the next number that has not been crossed out and circle it in red and continue in the same way.
- After they have done the top row, students may realise that they can cross out all numbers in the columns containing even numbers and multiples of 5 and 10.

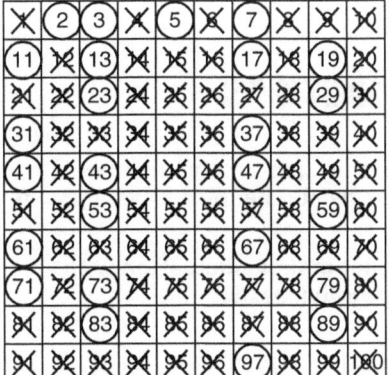

Students then take turns to read out, in order, the numbers that have red circles around them. The teacher writes them on the board and asks:

- What do these numbers have in common?
- Which digits appear as final digits for the prime numbers?
- Which digits do not appear at the end of prime numbers?

Finally, students use their own words to describe the activity and explain the key features of prime numbers.

Hunt multiple factors

'A factor is a number that divides exactly into another number; a multiple is the result of multiplying one number by another number.'

Quick recall of factors and multiples makes calculation with fractions and many areas of maths much easier.

Taking it further

A player gains two extra points every time they finish their turn on a prime number. Colour in the number.

This game for two players gives practice in finding factors and multiples of numbers. It can be played at two levels: the basic game or the strategic version.

Basic game

The playing board is a 100 square. Players take turns to find a factor or multiple of the last number calculated by their opponent.

- Players keep track of all the moves in a table (see example below).
- Player A chooses a number below 20 as a starting number and circles it.
- Player B finds a factor or multiple of Player A's number and crosses it out.
- Player A finds a factor or multiple of Player B's number and circles it.
- Play continues until one player cannot execute a move.
- The winner is the last person to circle or cross out a number.

A game in progress:

MOVES	
Player A – CIRCLE	**Player B – CROSS**
Start 14	2 (factor of 14)
8 (multiple of 2)	64 (multiple of 8)
32 (factor of 64)	4 (factor of 32)
28 (multiple of 4)	7 (factor of 28)

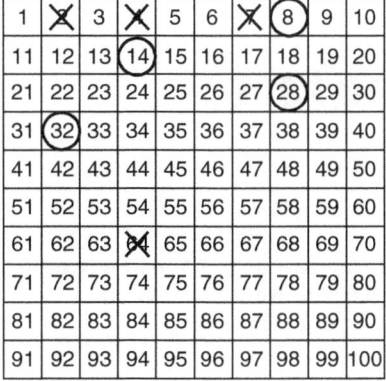

1	✗	3	✗	5	6	✗	⑧	9	10
11	12	13	⑭	15	16	17	18	19	20
21	22	23	24	25	26	27	㉘	29	30
31	㉜	33	34	35	36	37	38	39	40
41	42	43	44	45	46	47	48	49	50
51	52	53	54	55	56	57	58	59	60
61	62	63	✗	65	66	67	68	69	70
71	72	73	74	75	76	77	78	79	80
81	82	83	84	85	86	87	88	89	90
91	92	93	94	95	96	97	98	99	100

Bonus idea ★

Investigate with students how the initial starting number affects the course of the game:

- **Which number results in the longest chain of moves?**
- **Which number, or numbers, results in the shortest number of moves?**
- **How do prime numbers affect play?**
- **Do chains arising from square numbers give an advantage or not?**

Strategy game

- Moves are made in the same way as for the basic game.
- Now players have two winning strategies to pursue: three in a row or preventing the opponent from moving. This requires seeing certain numbers as target numbers, either to create a line of three or to sabotage the opponent's potential for a line of three.
- The winner is the first person to cross out three in a row (horizontally, vertically or diagonally) or to stop their opponent moving.

With thanks to NRICH for this idea.
See https://nrich.maths.org/5468.

60-second prime factors

'Challenge students to find the prime factors of a given number in 60 seconds.'

Every whole number that is not prime can be written as a product of its prime factors. They are the basic building blocks for encryption for security codes and computer coding.

Students use a factor tree and chain multiplication to find all the prime factors of a number. Check that students know that a prime number has only two factors: itself and 1. 1 is not a prime number because it has only one factor: itself.

The factor tree

The name 'factor tree' describes the branching structure of the diagram. Work through the following example together, then ask students to factorise other numbers.

- Write 60 and draw two lines like an inverted 'V'.
- Choose any two factors and write them on the diagram as shown.
- Repeat the process until all the factors are prime factors.
- Circle the prime factors to identify them.
- Write the prime factors as an equation using chain multiplication and then power notation.

Factor trees can be drawn in the following two ways. Whichever way they are drawn the result is the same. The first example suggests a spatial visualisation whilst the second follows a more sequential format. Students choose the representation they prefer.

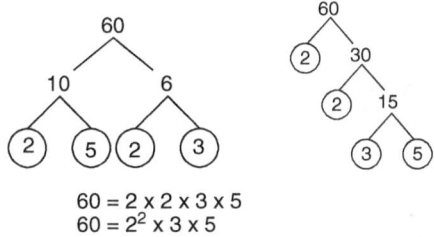

$$60 = 2 \times 2 \times 3 \times 5$$
$$60 = 2^2 \times 3 \times 5$$

Taking it further

Students can colour code the prime factors to make it easier to identify how many there are of each prime number.

Step-by-step alternative

Some students find the factor tree visually confusing. They can try this step-by-step approach instead:

- Break 60 into any two factors. 10 is a good starting point because it is the product of two prime factors.
- Use tree notation as shown to find the factors of 10.
- Write the new equation as shown.
- Continue until all the factors are prime factors.
- Write the equation in power notation.

$$60 = 10 \times 6$$
$$60 = 2 \times 5 \times 6$$
$$60 = 2 \times 5 \times 2 \times 3$$

$$60 = 2^2 \times 3 \times 5$$

Square numbers

'Why is a number multiplied by itself called a square?'

Teach square and cube numbers as geometric shapes to make them memorable. Being able to quickly recognise them is a handy tool for quick estimates in calculations.

A square number is a number multiplied by itself. It is called a square number because the answer can be modelled as a square. This process produces a geometric progression, which is a concept inherent in ideas of scaling and has many practical applications. Being comfortable working with squares lays the foundations for work with powers and standard form (see Part 5).

- Use base ten equipment to model the squares of each of the following numbers: 2, 3, 4, 5.
- Draw a diagram of each model on 1 cm^2 paper, e.g.:

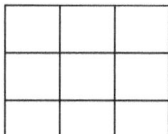

$$3^2 = 9$$

- Record the result in a table using both square notation and the expanded form to emphasise the meaning. Use n to represent any number.

n	$n \times n = n^2$	n^2 (square number)
2	$2 \times 2 = 2^2$	$2^2 = 4$
3	$3 \times 3 = 3^2$	$3^2 = 9$
4	$4 \times 4 = 4^2$	$4^2 = 16$
5	$5 \times 5 = 5^2$	$5^2 = 25$

Cryptography – crack the code

'This idea shows why prime factors are important and provides practice in long multiplication.'

Prime numbers are essential to encryption, which is the basis of internet security. A semi-prime is the result of multiplying two prime numbers together.

In this activity, students multiply prime numbers to generate a secret message. Then, they decipher the message by using the more familiar number-to-letter code substitution.

The message consists of a three-letter word and a four-letter word. All the letters are different. Students obtain the encrypted message by finding suitable semi-prime numbers, then use the code to substitute letters for numbers.

- Students work in small groups to share the task of finding the encrypted message.
- Write this list of prime numbers on the board: 13, 23, 41, 73, 97. Students use them to generate all the semi-prime numbers they can. (There will be 10 different numbers.)
- Choose a three-digit and a four-digit number in which all the digits are different. None of the seven digits should be the same.
- Now the teacher writes the number-to-letter code on the board.
- Students substitute letters for numbers to find the mystery message.

Example:

23 x 41 = 943 and 73 x 97 = 7,081

0	1	2	3	4	5	6	7	8	9
i	e	s	n	u	p	r	t	m	f

The message is: fun time.

Fractions, decimals and percentages

Part 3

Fraction foundations

'It's easy to see why the denominator gets bigger when the fraction gets smaller.'

Fold paper strips to make the meaning of fractions clear: one whole is divided into equal parts.

Provide strips of paper that are all the same length. Hold up a single strip (horizontally) and say, 'This is one whole. Can you fold it in half?' Students then fold the paper in half and draw a line to emphasise the fold line. Discuss with the students how many parts there are and ask them to shade one part.

Then, the teacher folds another strip into two unequal parts and shades one part. For example:

Display it and ask, 'How many parts are there? How many parts are shaded? Is the strip folded in half?' Students discuss the difference between the two parts and between two equal parts.

Explain that a fraction is written as two numbers, one above the other, separated by a line. The bottom number (denominator) shows how many parts the whole is divided into. The top number (numerator) tells you how many parts are shaded.

Ask students to paste the first paper strip into a book and write $\frac{1}{2}$ on the shaded part.

Investigate the effect of folding a strip into half and half again. Students shade one part, write the fraction and paste it under the first strip.

$\frac{1}{2}$	

$\frac{1}{4}$			

Fraction bars and lines

'The students began to understand that doubling the number in the denominator results in a line only half as long.'

The skill of quickly judging the comparative size of fractions on lines is crucial to working with graphs.

Help students to depict and interpret unit fractions on lines to show that the smaller the denominator the larger the fraction – or the larger the denominator the smaller the fraction.

Linear fractions

- Model fractions using a paper strip (Idea 20). Start with $\frac{1}{2}$ and ask students to paste the fraction into their books.
- Students then draw a line the same length as the paper strip. Note the small marks to indicate the beginning and end of the line. They should mark the halfway point and use an arc to emphasise half. Make sure they annotate the diagram.

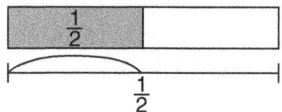

- Repeat the procedure for: $\frac{1}{4}$ $\frac{1}{8}$ $\frac{1}{16}$

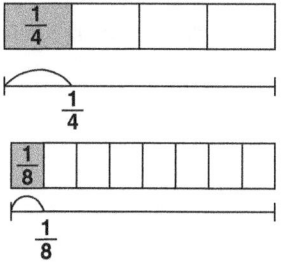

Teaching tip

It is important that students draw their lines freehand and do not use a ruler to measure the distance. This places the emphasis on the relationship between the quantities. It also helps develop crucial skills such as visualisation and approximation.

Taking it further

Supply this list of unit fractions in random order:

$$\frac{1}{12} \quad \frac{1}{5} \quad \frac{1}{8} \quad \frac{1}{3} \quad \frac{1}{2}$$

Students write the fractions in order, starting with the smallest fraction, and use inequality signs to show the relationship. Students locate the fractions on a number line to check their answers.

Fractions: the area model

'I could actually see straight away which fractions were bigger, rather than having to imagine it in my head.'

Quickly judging the comparative size of fractions and interpreting proportions is a life skill. Set the scene for interpreting graphical information by building visual images of fractions of shapes.

Rectangles provide an area model of fractions. This work provides essential foundations for geometric and algebraic calculations. They help students understand that the phrase 'the bigger the denominator the smaller the fraction' means that as the denominator increases, each portion becomes smaller.

Explore the area model of fractions by folding paper rectangles in two dimensions – vertically and horizontally.

- Students fold the paper in half horizontally and vertically as shown in the model.
- Model the fractions $\frac{3}{4}$ $\frac{4}{6}$ and $\frac{5}{8}$ by folding paper rectangles.
- Draw diagrams and colour them in to show the size of each fraction. Put them in order of size.

$\frac{5}{8}$ 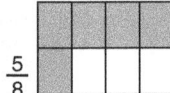 $\frac{4}{6}$ 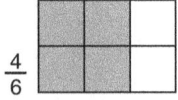 $\frac{3}{4}$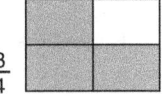

- Students work in pairs to discuss the meaning of each diagram.

Linear fractions

- Write the fractions that have been modelled in order of size starting with the smallest fraction: $\frac{5}{8}$ $\frac{4}{6}$ $\frac{3}{4}$
- Draw the fractions on lines to compare them and reinforce the visual proportional representation.

Taking it further

This model can also be used to demonstrate that multiplying fractions makes the product smaller, while dividing fractions leads to a larger answer (see Ideas 36 and 37). This is confusing for students who believe that multiplying always makes a quantity bigger, while dividing makes it smaller.

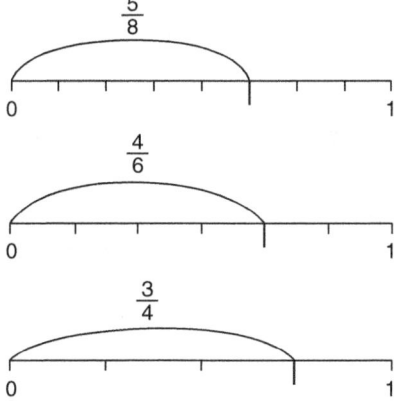

Numerator and denominator effects

'My students are much more confident about the role of the top and bottom numbers in a fraction.'

Change the numerator (top number) or the denominator (bottom number) to investigate their roles in a fraction and the relationship between them.

If students have difficulty drawing accurate diagrams for the 'numerator change' activity, provide each student with a base sheet with circles divided into fractions. (See online resources.)

This activity also introduces the idea of 'top-heavy' fractions, which represent more than one whole, and mixed numbers, which contain a whole number and a fraction. Discuss with students how to depict a fraction with a numerator that is larger than the denominator. This involves one (or more) whole shapes split into fractions as well as the portion of the whole.

Compare fractions to investigate the different functions of the denominator and the numerator. Create and record a series of fractions with students by keeping one number in the fraction constant and changing the other one.

Numerator change and denominator constant

- Choose the constant denominator.
- Students draw circles divided into the designated number of parts.
- Students take turns to roll a 0–9 dice to generate a number for the numerator. If a zero is thrown, the player rolls again. If they roll a number that is bigger than the denominator, the player rolls again.
- The player writes the fraction and colours in the fraction on the circle.
- The winner is the person who finds four consecutive fractions.
- They must sketch them on a number line to prove they haven't left any out!

Example of a winning sequence with denominator 8:

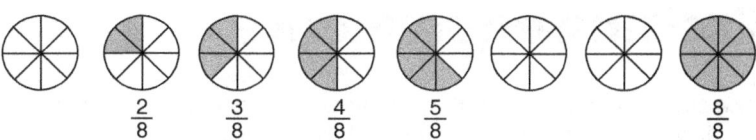

$\frac{2}{8}$ $\frac{3}{8}$ $\frac{4}{8}$ $\frac{5}{8}$ $\frac{8}{8}$

Winning sequence on a number line:

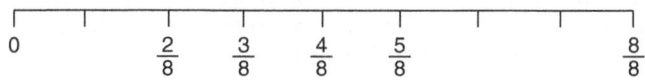

0 $\frac{2}{8}$ $\frac{3}{8}$ $\frac{4}{8}$ $\frac{5}{8}$ $\frac{8}{8}$

Denominator change and numerator constant

- Provide a base sheet of rectangles. It is important that the rectangles are the same size. (See online resources.)
- Choose the constant numerator.
- Students take turns to roll a 0–9 dice to generate a number for the denominator. If a zero is thrown, the player rolls again.
- The player writes the fraction and divides the rectangle to show the denominator.
- They colour in the portions to represent the numerator.
- The winner is the person who finds four consecutive fractions.

Example of a winning sequence with numerator 4:

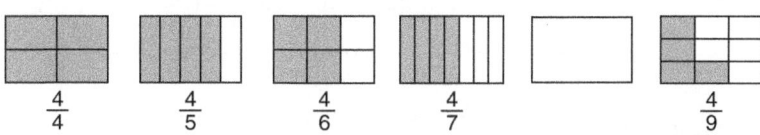

$\frac{4}{4}$ $\frac{4}{5}$ $\frac{4}{6}$ $\frac{4}{7}$ $\frac{4}{9}$

Exciting equivalence

'The students learned that they could create effectively the same fraction expressed across a range of base values.'

Equivalent fractions are fractions that are of equal value. Work with equivalence develops understanding of the proportional relationship within and between fractions.

Teaching tip

Encourage students to use equipment – Cuisenaire rods or paper strips – if they are unsure of the process. Students need to be confident finding equivalent fractions in order to compare them, to add them or to subtract them.

Equivalent investigation

Investigate with students the following question: 'How many eighths are equivalent to three quarters?'

- Use Cuisenaire rods to model $\frac{3}{4}$ next to an empty line marked only with 0 and 1.

- Place rods, each worth $\frac{1}{8}$ of the line, to find out how many are needed to be the same length as $\frac{3}{4}$

- Draw a diagram and write an equation to show the relationship:

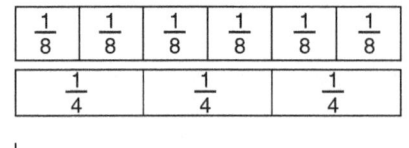

$$\frac{3}{4} = \frac{6}{8}$$

- Investigate other equivalences:

$$\frac{2}{3} = \frac{?}{9} \qquad \frac{3}{5} = \frac{?}{20} \qquad \frac{21}{28} = \frac{?}{4}$$

Equivalent shapes

Provide a worksheet showing a variety of fractions of shapes (see online resources). Ask students to:

- Write the fraction of each shape.
- Reduce the fraction to the simplest form.
- Create their own examples.

Sequence solutions

Write a sequence of fractions on the board and ask students to copy it into their books then complete the sequence. Encourage students to show their workings by using diagrams or number lines. You could use the following sequences:

$$\frac{1}{2} = \frac{2}{4} = \frac{3}{?} = \frac{4}{?} = \frac{5}{?}$$

$$\frac{1}{3} = \frac{2}{6} = \frac{3}{?} = \frac{4}{?} = \frac{?}{?}$$

$$\frac{1}{4} = \frac{2}{8} = \frac{3}{?} = \frac{4}{?} = \frac{?}{?}$$

Teaching tip

Students need to discuss what they have done using their own words. At first the explanations may be cumbersome but the descriptions will become clearer as they develop their understanding.

41

Fraction comparison calculations

'The method helped me compare fractions that I was struggling to rank just by looking at them.'

A common denominator makes it easy to compare the size of fractions (see Idea 23). In this activity, students find the lowest common denominator (LCD) of the fractions and then turn them into equivalent fractions.

The process of finding the LCD is often taught as a rule; however, it is best if students reason through the steps. This not only helps them understand the meaning of the calculation but also lays important foundations for working with algebraic equations.

With the students, compare $\frac{2}{3}$ and $\frac{3}{4}$ Which is bigger?

- Look at the denominators and ask students whether they have a common factor. The answer is no, they do not.
- Find the smallest number that has both 3 and 4 as a factor. This is called the lowest common multiple (LCM). To do this, draw up a table with multiples of 3 and multiples of 4:

Multiples of 3	3 6 9 ⑫
Multiples of 4	4 8 ⑫

- 12 is the lowest number that has 3 and 4 as factors. 12 = 3 x 4. This gives a useful rule for finding a common denominator: multiply the denominators by each other to find a common denominator. (Note that it might not be the LCD.)
- Convert each fraction into an equivalent fraction with the denominator of 12.

- Discuss the process and record it in a flow
  diagram:

   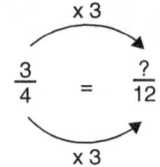

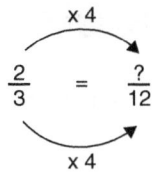

$$\frac{2}{3} = \frac{?}{12}$$ (×4 top, ×4 bottom)

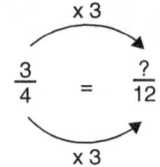

$$\frac{3}{4} = \frac{?}{12}$$ (×3 top, ×3 bottom)

Taking it further

Make sure students can explain what they are doing and why. They will then be able to adapt this method for working with ratio and proportion (see Part 6).

- Write the equation to show the steps in the
  calculation:

$$\frac{2 \times 4}{3 \times 4} = \frac{8}{12}$$
 $$\frac{3 \times 3}{4 \times 3} = \frac{9}{12}$$

- The question asked which fraction is bigger.
- Write the answer as an inequality: $\frac{9}{12} > \frac{8}{12}$
- Therefore: $\frac{3}{4} > \frac{2}{3}$

Simplifying fraction models

'I now understand what it means to express a fraction in its lowest terms.'

Help students to put a fraction in its simplest terms to make it easier to understand.

Teaching tip

Modelling large numbers with unit cubes may seem tedious; however, it is worth doing in the initial stages as it develops a visual image that helps establish relative size of quantities.

Use models and diagrams to show how much easier it is to imagine a simplified amount. Make it clear that reducing a fraction to the lowest terms changes the numbers but it does not change the value of the fraction.

Demonstrate the effect of simplifying fractions using diagrams, base ten equipment, Cuisenaire rods and number lines:

- Provide a diagram showing the fraction $\frac{21}{28}$ as a line of squares.
- Students use base ten unit cubes to model the fraction $\frac{21}{28}$.
- Ask, 'Can you model an easier way to make this fraction easier to imagine?'
- Students use Cuisenaire rods to experiment to find that $\frac{3}{4}$ is equivalent to $\frac{21}{28}$.

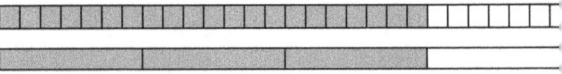

- Now show the fractions on a number line:

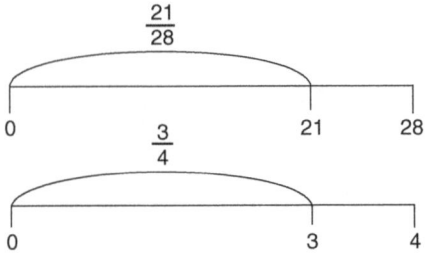

- Write out the calculation showing the simplification:

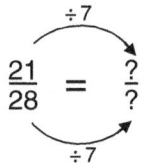

$$\frac{21 \div 7}{28 \div 7} = \frac{3}{4}$$

Now ask students to simplify the following fractions. They should model them using base ten unit cubes and Cuisenaire rods, show them on number lines and write the calculation:

$$\frac{25}{60} \qquad \frac{12}{15} \qquad \frac{56}{64} \qquad \frac{27}{45}$$

Taking it further

Rank fractions in order of size starting with the smallest and using the inequality symbol.

Fraction wall

'The wall makes it so easy to see why one third is bigger than one quarter.'

A fraction wall makes a dramatic image of the relationships between fractions. It is easy to compare the size of fractions to find equivalent amounts and common denominators.

Taking it further

Discuss how a fraction wall can also be used as a factor wall. For example, if the one whole is expressed as a length, then each 'brick' represents a length that is a factor.

Fraction wall

Students use Cuisenaire rods to build a fraction wall. This consists of layers of different 'bricks'. The bricks within each layer are the same size. The width of each layer is one whole unit.

- Students work in pairs to build a fraction wall and explore all the ways of dividing one whole into equal-sized parts.
- Provide each student with a bar cut out of card. The bar should measure 12 cm but do not tell them how long it is. This avoids the problem of students fixating on the measurements rather than focusing on the proportional relationships.
- Allow students to start with any size rod they want and gradually derive the fractions shown in the diagram.
- Each student should then draw an accurate diagram of the fraction wall.

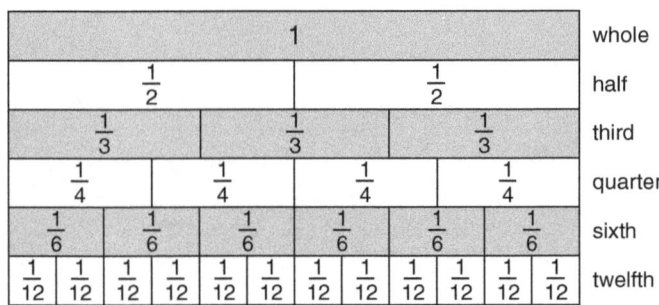

Fraction families

Students use Cuisenaire rods and fraction walls to find the equivalent fractions within the following numbers: 10, 15, 18.

1										whole
$\frac{1}{2}$					$\frac{1}{2}$					half
$\frac{1}{5}$		$\frac{1}{5}$		$\frac{1}{5}$		$\frac{1}{5}$		$\frac{1}{5}$		fifth
$\frac{1}{10}$	$\frac{1}{10}$	$\frac{1}{10}$	$\frac{1}{10}$	$\frac{1}{10}$	$\frac{1}{10}$	$\frac{1}{10}$	$\frac{1}{10}$	$\frac{1}{10}$	$\frac{1}{10}$	tenth

Wall charts of time fractions

Students find the fractions of some numbers associated with time: 12, 24, 60. To do this, students make paper displays showing the fractions of large numbers. They do not measure any of the lengths of the pieces of paper. In constructing the fractions by folding paper, they focus on the relative size and so develop a better understanding.

Taking it further

Draw a number line and use arcs to show the halves and quarters above the line, and thirds and sixths below the line. Use different colours for each series of fractions.

Fractions of sets

'So, what is a "whole" anyway?'

Students need to understand the concept of fractions as part of a set. Young children learn that a fraction is part of one whole; however, this is a limited construct and leads to later misconceptions. Use objects to model a set as a collection of objects.

Use money to explore the relationship between different categories within a larger group. Here there are several possible criteria of comparison: the size of the coins, the number of sides of the coins or the value of the coins.

Students work in pairs as they need to talk about what they are doing and express their ideas in diagrams. They consider part of a group of objects in relation to the whole group. First, establish the criteria of comparison. This is crucial for classification and some students find it very difficult.

- Provide a container of counters of two colours, say black and white.
- Students each take a handful of counters and put them together.
- They count the total number of counters and write it down. This is called the set of objects in the collection.
- Encourage students to discuss the criteria of comparison. In this case there is only one possibility – colour.
- The students draw a large circle to show the set of counters. They draw a smaller circle inside to show a subset of counters.

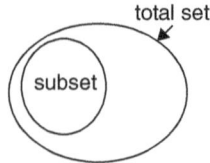

- They put all the counters into the circle, then sort all the counters of one colour (black in the example) into the smaller circle. The students discuss how many counters are in the whole set and how many counters are in the subset. They then summarise the information in a table.
- They write the quantity of black counters as a fraction of the collection. The total set of objects is the denominator (bottom number) and the subset is the numerator (top number). They then simplify the fraction and write a one-sentence summary.

Example: What is the fraction of black counters?

Total number of counters 12

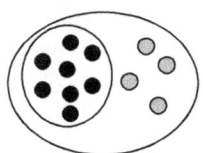

| Black counters | 8 |
| White counters | 4 |

Fraction of black counters $\frac{8}{12} = \frac{2}{3}$

$\frac{2}{3}$ of the counters are black.

Bonus idea ★

Many students find fractions difficult because they are not clear about what constitutes the whole group and what is part of the group. Broaden the topic and discuss how things are identified, labelled and categorised according to underlying principles. Explore how this is used to classify species in biology.

Percentages

'The percentage sign is a quick way of writing "out of 100". It has the zeros divided by a line so I think of it as 100 written in an interesting way: %.'

A percentage is the number of parts per hundred. Percentages matter because they make it possible to compare numerical information. A percentage is simply a fraction with a denominator (bottom number) of 100.

Teaching tip

This exercise develops a strong visual image of the relationship between fractions and percentages. This is invaluable for calculations with rates of change. (See Ideas 57 and 58.)

Many students struggle with percentages but they are easy to use if they are modelled on a 100 square and directly related to equivalent fractions. 'Per cent' means 'out of 100'. Help students to visualise percentages by modelling equivalent fractions with base ten material.

Provide each student with a square 10 cm x 10 cm but do not tell them the dimensions or allow them to measure them.

Students divide the square into quarters and shade in $\frac{1}{4}$

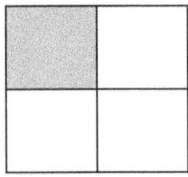

They fill the shaded portion with 1 cm² blocks. Ask, 'How many blocks are required?' (25)

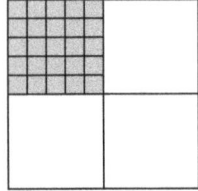

Discuss with the students how to use equivalent fractions to work out how many 1 cm² blocks are required to fill the whole square. Together, write the calculation as a process (function diagram). Use the letter N to denote the unknown quantity.

Taking it further

Explore a linear model of percentages. Use a metre stick to indicate 100% and display the percentage using base ten rods and unit cubs in a line next to it.

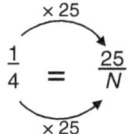

$$\frac{1}{4} = \frac{25}{N}$$

with $\times 25$ arrows

Now, write the fraction as an equation with N as the subject. Use the properties of equality and show all the stages of working:

$$\frac{1}{4} = \frac{25}{N}$$

$$N \times \frac{1}{4} = \frac{25}{N} \times N$$

Multiply both sides by N. Cancel where appropriate.

$$N \times \frac{1}{N} \times 4 = 25 \times 4$$

Isolate the unknown by multiplying both sides by 4 and cancelling.

$$N = 100$$

Finally, express 25 as a fraction of 100:

$$\frac{25}{100} \text{ is } 25\%$$

Now model these percentages on a 100 square and ask students to write the percentage as an equivalent fraction in its simplest form: 50%, 75%, 30%, 60%

Decimals

'The students were soon able to understand that numbers with decimals, which they see in everyday life, are just fractions expressed in a different form.'

Decimals are fractions expressed as part of the base ten place value system. The decimal point marks the boundary between whole numbers and parts of whole numbers. A decimal can be expressed as a fraction with a denominator of 10 or multiple of 10.

Teaching tip

Students need to fold the strip of paper so that the portions are of equal size. The best way is to fold it in half and then in a concertina fashion. If they start at one end and fold one portion over the other (like rolling a carpet), this quickly distorts the proportions.

Make sure that students understand the concept of a decimal fraction to avoid any misunderstandings. Use a number line to link decimals to fractions with a denominator of 10 (or multiple of 10) to make the fractional meaning clear.

Taming the tenths

- Model tenths using a paper strip (Idea 20). Give students a plain paper strip and encourage them to try folding it into tenths. It is very difficult to achieve this with reasonable accuracy so if necessary provide photocopied strips with the lines for the folds shown. (See online resources.)
- Ask students to colour in one portion and write $\frac{1}{10}$ in each portion.
- Students then stick the paper strip onto a piece of paper.
- They draw a line the same length as the paper strip and mark each tenth position on the line.
- Then, ask them to mark the following fractions on the number line: $\frac{3}{10}$ $\frac{5}{10}$ $\frac{8}{10}$

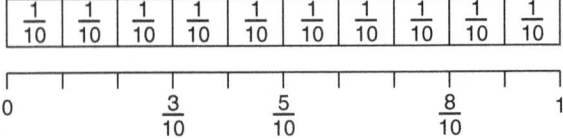

- With the students, draw a place value grid and express the fractions as decimals on the grid. Discuss the ways of referring to fractions and decimals.

Taking it further

Extend the game 'Playing with hundredths'. Link the value of decimals on the place value grid to the number line by recording each move on a number line as consecutive arcs. This emphasises the comparative size of each decimal value.

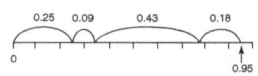

	HTU $\bullet \frac{1}{10}$	
$\frac{1}{10}$ one tenth	0 $\bullet$ 1	point one
$\frac{3}{10}$ three tenths	0 $\bullet$ 3	point three
$\frac{5}{10}$ five tenths	0 $\bullet$ 5	point five
$\frac{8}{10}$ eight tenths	0 $\bullet$ 8	point eight

Comparing decimals and fractions

Students practise writing fractions as decimals and decimals as fractions.

- Ask students to write these fractions as decimals and locate them on a number line:
  $\frac{5}{10}$ $\frac{1}{5}$ $\frac{2}{5}$
- Students sketch a new number line for each decimal, estimate the position of the decimal and mark it on the number line.
- They draw an arc to show the proportion of the line and write the fraction it represents.

Playing with hundredths

- Students practise adding decimals with two decimal places (2 d.p.) and show them on the place value grid.
- Players take turns to roll two 0–9 dice and make a number with two decimal places (2 d.p).
- Each player must take four turns.
- They write each number under hundreds, tens, units (HTU) headings.
- They add the numbers.
- The winner is the person who has the total closest to 1.

```
H T U • 1/10  1/100
    0 •  2   5
    0 •  0   9
    0 •  4   3
    0 •  1   8 +
    0 •  9   5
```

Same fraction, different form

'I learned that you can take any number that is a fraction and also write it as a decimal number or a percentage, and it is the same actual amount.'

Use number lines to show equivalence between fractions, decimals and percentages. They are different ways of representing the same idea – how something can be divided up or shared out. The key concept in all of them is proportion.

Teaching tip

The ability to quickly sketch and read a number line is one of the most useful tools in the mathematics for life toolbox.

Students need to be able to convert any given fraction, percentage or decimal to each of the other systems. Converting from one form of fraction to another is often taught as a rule that many students struggle to remember, so start by developing reasoning on a number line. This forms a solid base with a memorable image to draw on before learning the calculations.

Location practice

- Draw three lines of the same length. Remember that equivalent fractions have the same proportions.
- Discuss how to label the end points of the lines and why. At this stage students are investigating fractions as part of one whole so two of the lines will represent the distance 0 to 1. Percentages are out of 100 so the third line will represent 0 to 100.
- Reason to locate the fraction $\frac{1}{4}$ on a line. Instil the habit of always identifying the midpoint as the first step.
- Mark $\frac{1}{4}$ as a decimal or percentage in the equivalent positions on the other number lines.

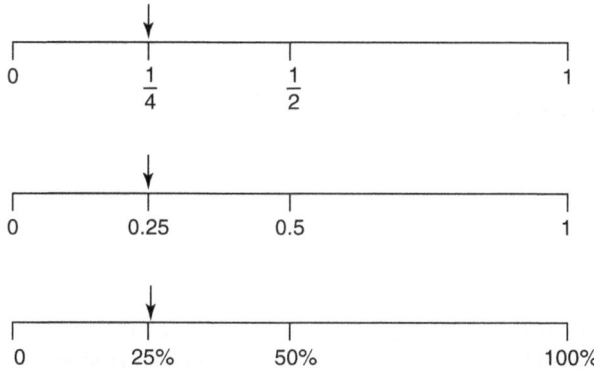

Now ask students to use number lines to illustrate the conversions necessary to complete this equivalent fraction table:

Fraction	Decimal	Percentage
	0.4	
		$12\frac{1}{2}\%$
$\frac{3}{5}$		

Bonus idea ★

Learn to recognise equivalent fraction forms by matching percentage, fraction and decimal cards in the memory game (see online resources).

Four fractions to make a whole

'Have fun converting between the different ways to represent fractions.'

Help students learn to quickly judge the size of a fraction in relation to other fractions and to move easily between fractions, decimals and percentages.

Taking it further

Students play the same game with four dice and a target of 10 to practise conversions between mixed numbers.

This game reinforces the link between fractions, decimals and percentages as different ways of expressing the same amount. Being able to quickly convert between the different representations helps to select the most appropriate form for calculations. Decimals are required for calculator work but fractions may be easier to use in mental calculations. Quick sketches on number lines help to work out rough estimates.

Grid game

- The aim of the game is to gain a total score as close to 1 as possible after four throws.
- Students play in groups of up to four players. Each player draws up their own score sheet consisting of three columns headed 'Fraction', 'Decimal' and 'Percentage'.
- The players take turns to roll two 0–9 dice and make a number that is a decimal, a fraction or a percentage. If a player rolls zero, they roll the dice again.
- The player records their chosen number in red on the score sheet.
- They write the equivalent amounts in the other two columns, simplifying the fractions and rounding to two significant figures if necessary (there may be more than two decimal places when fractions are converted to decimals).

- After four rounds, players add their total amount in the decimal column. They check the answer by adding the total in each column. The winner is the person whose total is closest to 1.

Example of Player A

(The chosen number is shown in bold.)

Numbers on dice	Decimal	Fraction in simplest form	Percentage
4, 7	**0.47**	$\frac{47}{100}$	47%
8, 2	0.25	$\frac{2}{8} = \frac{1}{4}$	25%
3, 9	0.33	$\frac{3}{9} = \frac{1}{3}$	$33\frac{1}{3}$%
6, 1	**0.16**	$\frac{16}{100} = \frac{4}{25}$	16%
Total score	1.21	$1\frac{21}{100}$	$121\frac{1}{3}$%

> **Bonus idea** ★
>
> Students can play the same game on an empty number line. Students record each throw on the number line. This is more difficult as it requires judgement as well as accuracy.

Memorable fraction forms

'I enjoyed this game; it made me think about different ways of expressing fractions.'

Fractions, decimals and percentages are everywhere in our lives. It's important that students learn to quickly convert between them.

Match decimals, fractions and percentages in this memory card game for up to four players.

- You will need a set of cards for decimals, fractions and percentages. Each set is a different colour. (See online resources.) Some students find it very confusing to play with three sets of cards so simplify the game by playing with two sets of cards initially.
- Shuffle two packs of cards and turn them face down on the table in a structured arrangement. This helps students with poor spatial skills access the game.
- Players take turns to turn up a card, read it aloud and then say what they are looking for, e.g. Player A turns up $\frac{1}{4}$ and says, 'One quarter. I am looking for 25%.'
- The player then selects a second card. If it is the equivalent amount, the player keeps the cards and has another turn. If the cards do not match, the player explains why, e.g. Player A turns up $\frac{1}{4}$ and 50%, and says, '50 per cent is $\frac{1}{2}$ I needed 25 per cent to match $\frac{1}{4}$.'
- The winner is the player with the most cards when they have all been paired up.
- Once students are confident with finding pairs, play with all three packs of cards. A winning move requires three matching cards.
- All players fill in an equivalent fraction table as the cards are identified (see Idea 31). There is an example in the online resources.

Improper and mixed fractions

'I used to think that fractions were always smaller than 1, but I now know that any number can be a fraction, even the big ones!'

Some students think that a fraction must be less than 1 so that the numerator (top number) is always smaller than the denominator (bottom number). This is not true. A fraction in which the top number is larger than the bottom number is called an improper fraction. This can be expressed as a mixed number that has an integer part and a fraction part.

Model and describe improper fractions and mixed numbers to demonstrate that there are different ways of describing the same quantity. Provide strips of paper of the same length and width and then model $\frac{7}{4}$ by folding paper with the students.

- Take a strip of paper, divide it into quarters and mark the quarters.
  - How many quarters are there? [4 quarters]
  - How many quarters are required? [7 quarters]
  - There are 4 quarters. How many more quarters are required? [3 more quarters]
  - How many strips of paper are needed? [2 strips of the same length]
- Fold another strip of paper into quarters.
- Stick both pieces of paper into a book and colour in $\frac{7}{4}$

- Draw a line the same length as the paper strips. Indicate the beginning and end of each paper strip on the line and mark the points 0, 1 and 2.
- Divide the line into quarters.
- Use arcs to show that $\frac{7}{4}$ is $\frac{4}{4}$ and $\frac{3}{4}$
- Since $\frac{4}{4} = 1$, you can write $\frac{7}{4}$ as $1\frac{3}{4}$

> **Teaching tip**
>
> Don't rush. Encourage students to ask questions to help clarify what to do and give them plenty of practice. After the whole-class modelling activity, ask them to work in pairs and use two dice to generate numbers to express as improper fractions and mixed numbers following the same process.

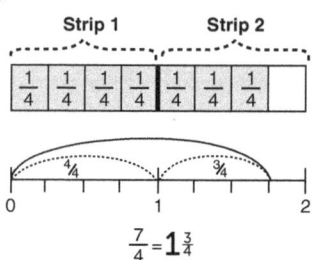

$$\frac{7}{4} = 1\frac{3}{4}$$

Adding and subtracting fractions

'All the work on finding equivalent fractions makes sense now I can apply it to calculations.'

Avoid confusion with fraction calculations by using paper-folding and number lines to provide memorable images. It is vital students know what the segments represent.

Fractions in addition and subtraction tasks must have a common denominator (bottom number). That means that each part, or segment, will be the same size.

With the students, calculate $\frac{1}{2} + \frac{2}{5}$

- Find the LCD (lowest common denominator) and express both fractions in terms of the common denominator.
- The LCD is 10 so $\frac{1}{2} = \frac{5}{10}$ and $\frac{2}{5} = \frac{4}{10}$ (See Idea 25.)
- Model each fraction on a separate paper strip and colour them in.
- Draw the fractions on a number line to show that $\frac{5}{10} + \frac{4}{10} = \frac{9}{10}$
- Write the equation to show the original fractions and the fractions with the common denominator.

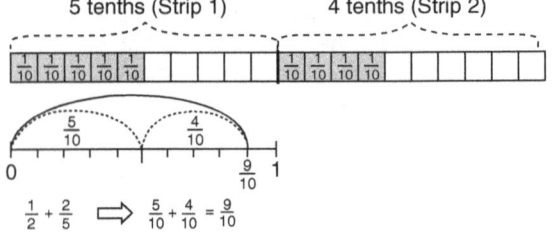

5 tenths (Strip 1)   4 tenths (Strip 2)

$\frac{1}{2} + \frac{2}{5} \implies \frac{5}{10} + \frac{4}{10} = \frac{9}{10}$

Now calculate $\frac{7}{8} - \frac{1}{2}$ with the students:

Taking it further

Give students plenty of practice so that they can confidently work with fraction calculations. These skills are involved in most areas of maths as well as physics and chemistry.

- The LCD is 8 so $\frac{7}{8}$ remains the same. $\frac{1}{2} = \frac{4}{8}$
- Model both fractions on a single paper strip.
- Shade the larger fraction in. Put crosses to indicate the amount subtracted.
- Show the subtraction on a number line.
- Write an expression and equation to summarise the process.

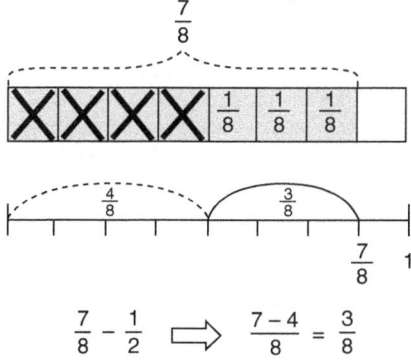

$\frac{7}{8} - \frac{1}{2} \implies \frac{7-4}{8} = \frac{3}{8}$

Multiplying by fractions

'The rules for multiplying and dividing fractions seem perverse as they appear to contradict common sense.'

Use flexible language to make it clear that the question to ask when multiplying fractions is: 'How many are there?'

Teaching tip

Students often find multiplication by fractions difficult if their only concept of multiplication is repeated addition. Revise the area model of multiplication. This method is an extension of the box method of multiplication (see online resources). Before starting, make sure students can fold paper accurately.

Multiply fractions and the answer is smaller than either of the terms; divide by a fraction and the answer is larger. However, all becomes clear if you focus on what the calculations mean, rather than simply finding an answer.

Start by exploring the multiplication of fractions (dividing fractions will be considered in Idea 37). Build on the paper folding activities in Idea 22 to model multiplication by a fraction as finding a portion of an area. This also explains why the term 'a fraction of an amount' has the same meaning as multiply by.

Use the area model to clarify the language and the concept of multiplication. With the students, calculate $\frac{1}{2} \times \frac{2}{3}$

- Fold a rectangular piece of paper into half and shade in one half.
- Fold the same piece of paper into thirds along the other axis.
- Use diagonal lines to identify two thirds.
- Draw a diagram of the model and mark the relevant fractions on the length of the sides as shown.

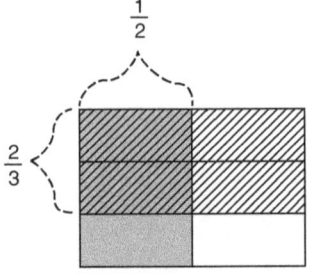

- How many segments are in the whole rectangle? [6 segments]
- What does this mean? [The rectangle is divided into sixths.]
- How many segments are part of both fractions? [2 segments, which is $\frac{2}{6}$ of the rectangle]
- Write the calculation to show all the steps involved:

$\frac{1}{2}$ of $\frac{2}{3} = \frac{1}{2} \times \frac{2}{3}$

$\frac{1 \times 2}{2 \times 3} = \frac{2}{6}$

$\frac{2}{6} = \frac{1}{3}$

Put numbers into contexts that make it easier to visualise the meaning.

Dividing by fractions

'The rule that to divide by a fraction you flip and multiply made no sense to me at all. Now I can see where it comes from.'

Put numbers in context to solve the conundrum that dividing by a fraction involves multiplication. Use flexible language to make it clear that the question to ask is: 'How many are there?'

The key words are 'in' and 'by'. They look so innocuous that many students don't really notice them. Consider 'divide *in* half' and 'divide *by* half'; they describe different concepts: sharing and grouping. This is not obvious from calculations with 'bare' numbers unless students describe what the results mean. Give students plenty of practice discussing models of division before introducing division by a fraction.

Pose questions to prompt discussion and model equations using counters, paper shapes or objects that can easily be cut, such as leaves.

Dividing in half

- Ask students, 'What does it mean to divide 10 in half?'
- Then ask, 'How do you write the equation?' [$10 \div 2 = ?$]
- Use counters to model ten counters split into two equal-sized groups. This is the sharing concept of division.
- Students explain the calculation in as varied a way as possible, e.g. 'I divided ten counters into two groups', 'I shared the counters equally so there are five in each group', 'Five is half of ten.'

    $10 \div 2 = 5$

- Now ask, 'Can you model $10 \div 2 = ?$ in a different way?'

- Students experiment and discuss the possibilities. Note that now they are not asked to find half of the amount.
- The equation can mean that ten is split into twos. Now, the question is: 'How many groups of two are there in ten?'
- Model groups of two to find that there will be five groups. This is the grouping concept of division.
- Describe the model: 'I wanted to find how many twos there are in ten. I put out pairs of counters and there were five groups. So, ten divided by two is five.'

$10 \div 2 = 5$

- Students work in pairs to make up stories that show how the sharing and the grouping concepts are different. They model the calculations and represent them in different ways, including diagrams, equations and fractions.
- Students share their ideas with the class. Encourage a variety of language.
- The teacher summarises the different ways of describing the results.

Dividing by a fraction

- Express division as grouping in a transparent way. Start with whole numbers, then introduce fractions:

$10 \div 2 = 5$	How many twos in 10?
$10 \div 5 = 2$	How many fives in 10?
$10 \div \frac{1}{2} = 20$	How many halves in 10?
$12 \div 3 = 4$	How many threes in 12?
$12 \div 4 = 3$	How many fours in 12?
$12 \div \frac{1}{3} = 36$	How many thirds in 12?

- Students then use fractions in contexts to illustrate dividing by a fraction to discover 'how many [fraction] in a quantity'. They model the calculations and represent them in different ways including diagrams, equations and fractions.

Taking it further

Give students plenty of practice so that they can confidently work with fraction calculations. These are life skills as well as being essential for economics and all the sciences.

Approximation, rounding and significant figures

Part 4

The rounding decider

'Decide on the decider then round a number up or down.'

Facility with rounding numbers is essential for work with significant figures. Students must establish the degree of accuracy required, then find the decider.

Taking it further

Students construct their own stories for rounding practice. This requires more focus on the relevance of numbers than answering prepared questions.

Rounding a number makes it easier to use. They are invaluable for quick estimates for calculations. Rounding a number up makes it slightly larger. Rounding a number down makes it slightly smaller.

Rounding practice

Round 5,926 with the students:

- Choose the appropriate place value position for the degree of accuracy required.
- The rounding number is the digit in the relevant place value position. Draw a line to the right of the rounding number.
- The digit to the right of the rounding number is the decider. If it is 5 or more, add one to the rounding digit. If the decider is 4 or less, leave the rounding digit unchanged.
- Round 5,926 to the nearest 10, then the nearest 100, and finally the nearest 1,000. Make the position of the rounding digit and the decider clear. Record the results in a table. Use the approximation sign $\approx$ to indicate that the number has been rounded.

Number	Rounded to nearest 10	Rounded to nearest 100	Rounded to nearest 1,000
5,926	592|6 $\approx$ 5,930	59|26 $\approx$ 5,900	5|926 $\approx$ 6,000

Now students can practise rounding four-digit numbers in pairs, using dice to generate the numbers.

Precision decisions

When students have grasped the concept of rounding, put numbers in contexts to investigate how accurate a number needs to be in practical situations. Students decide what level of accuracy is required and explain their answers. For example:

- How long will it take to walk to the station? The station is 27 minutes' walk away.
- How far is New York from London? A flight from London to New York is 5,585 km.
- How much does a laptop cost? The price on the ticket is £319.
- What is the fastest time anyone has run 100m? In 2009, Usain Bolt set a record time of 9.58 seconds.

The limits of rounding

'Pay attention to the language and symbolic notation that is used to describe the bounds of a rounded number.'

Start with a rounded number and find the range of values that it represents by identifying the lower bound and the upper bound. This is rounding in reverse.

Adopt a reasoning approach to establish the rule for calculating the bounds of a number. A rounded number is an approximation that lies midway between the lower and upper bounds. The largest value that the initial number can be is called the upper bound. The smallest value that the number can be is called the lower bound. The precise value will be in a range of values between the upper and lower bounds. This distance is known as the error interval.

Bounds on a number line

Students work in pairs to find and describe the bounds of various numbers by applying the problem-solving routine: state the question, identify relevant information and draw a diagram. Model this first:

- Pose the question: 'A number is rounded to the nearest 10. The rounded number is 40.' Ask:
  - What is the lowest number it can represent?
  - What is the highest number it can represent?
  - What information do we know? A number ending in 5 or above rounds up; a number ending in less than 5 rounds down.
- Draw a number line, marking 40 and using lines to show the lower and upper limits.
- Reason to establish that the lowest number must be 35 because 5 rounds up to 40. The upper limit is the highest value that would round down to 40. Can the upper bound be 45? Here there is a problem. Since 5 rounds

up to the next ten, 45 will round to 50. Therefore the upper bound must be less than 45. Does that mean the upper bound is 44?

- Discuss this question and use the number line to clarify reasoning. Mark 44 on the number line. This makes it clear that there are values between 44 and 45 that need to be taken into account.

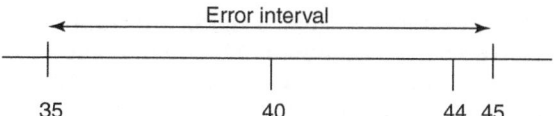

- Solve the dilemma by expressing the upper limit as less than 45.
- Describe the lower and upper bounds in words, then write the expression using the inequalities symbols (see Idea 12):

The lower bound is equal to or more than 35 and the upper bound is less than 45.

$$35 \leq 40 < 45$$

Quick fix for upper and lower bounds

You can find the upper and lower bounds by reasoning and using a number line. However, it is useful for students to be able to calculate them quickly. Model how to do this using the same question as above:

- Look at the degree of accuracy required (to the nearest 10 in the example).
- Divide the level of accuracy by 2.
- Add the result to the rounded number to find the upper bound. Subtract it to find the lower bound.
- Remember to use the inequalities symbols.

Example: 40 is rounded to the nearest 10. The degree of accuracy is 10.

$10 \div 2 = 5$
$40 - 5 = 35$
$40 + 5 = 45$
$35 \leq 40 < 45$

Bonus idea ★

Students work in pairs so that they can take turns to explain the process of finding the error interval and the upper and lower bounds of different values. Example questions:

- What are the lower and upper bounds of 34 km measured to the nearest 1 km?
- What are the lower and upper bounds of 760 grams measured to the nearest 100 grams?
- What are the lower and upper bounds of 280°C measured to the nearest 1°C?

Significant numbers

'My students found it surprising that the size of objects could affect their sense of the size of a number.'

Play around with objects of different scales and in different contexts to investigate how big a number is. Use equipment, diagrams, graphs and the imagination. Keep the numbers simple. Numbers rounded to a few significant figures are easier to understand.

Size is a relative concept: it needs to be compared to something else. Often students only judge numbers in relation to other numbers. The meaning of the value of a number is affected by its context. Our perception of size is shaped by both the quantity and the scale of the objects. These ideas may seem simplistic but are well worth exploring. Students are often surprised how their understanding of a number changes when they model it at different scales.

Students work in groups to model different-sized objects:

- Model 100 as an array using 1p coins.
- Model 100 as an array using grains of rice.
- Model 100 as an array of students sitting in rows in a hall.

The students photograph the arrays and discuss their reaction to the models.

Context consideration

Ask students to imagine 60,000. Invite the class to discuss whether this is a big number and why. Students then work in pairs to consider 60,000 in the following contexts. They summarise their reasoning and share ideas with the class.

60,000 spectators at a football match:

- How big is a football crowd?
- The capacity of the biggest stadium in the world is 150,000. The capacity of Wembley Stadium in London is 90,000.
- Is 60,000 a big number? Give reasons both for and against.

60,000 grains of wheat:

- How large a quantity is 60,000 grains of wheat?
- Relate grains of wheat to something on a human scale such as a loaf of bread. (A standard loaf of bread contains approximately 500 grams of wheat and 1 gram of wheat contains approximately 15 grains of wheat.)
- Is 60,000 a big number in this context?

Bonus idea ★

Wheat is a staple food. There are startling numbers on the Kansas wheat site to spark discussion on the global importance of this cereal we take for granted: http://kswheat. com/news/2015/07/22/ what-does-a-bushel-of-wheat-mean-to-me.

The Seven Summits

'Mountains don't change their size, yet rounding to significant figures makes it seem so.'

It is important to demonstrate to students that rounded numbers are approximations and need to be treated with caution.

Significant figures give rise to rounding errors. A rounding error is not a mistake; it is the difference between the rounded value and the actual value. It is important to remember this, as rounded numbers can give a false impression and judgements made using the rounded numbers can be wrong. Consider the bounds of the number when making judgements.

The Seven Summits are the highest peaks on each continent or continental area. Climbing all of them is a mountaineering challenge. Mountains are measured in both metric and imperial units:

Mountain	Height (ft)	Height (m)
Aconcagua	22,841	6,962
Carstensz Pyramid	16,023	4,884
Denali (Mount McKinley)	20,321	6,194
Elbrus	18,510	5,642
Everest	29,035	8,850
Kilimanjaro	19,340	5,895
Vinson	16,066	4,897

Students round each of the heights to two significant figures and record them in a table. They draw a bar chart to compare all the mountains in both metric and imperial units.

Discuss the results with the students and the rounding errors and misconceptions arising. What is the minimum number of significant figures needed to make a judgement? The misconceptions created by the rounding errors can be resolved by using number bounds. Ask the students to show the bounds for each rounded height.

Mountain heights to two significant figures (showing how simplifying numbers leads to rounding errors):

Mountain	Height (ft)	Height (m)
Aconcagua	23,000	7,000
Carstensz Pyramid	16,000	4,900
Denali (Mount McKinley)	20,000	6,200
Elbrus	19,000	5,600
Everest	29,000	8,900
Kilimanjaro	19,000	5,900
Vinson	16,000	4,900

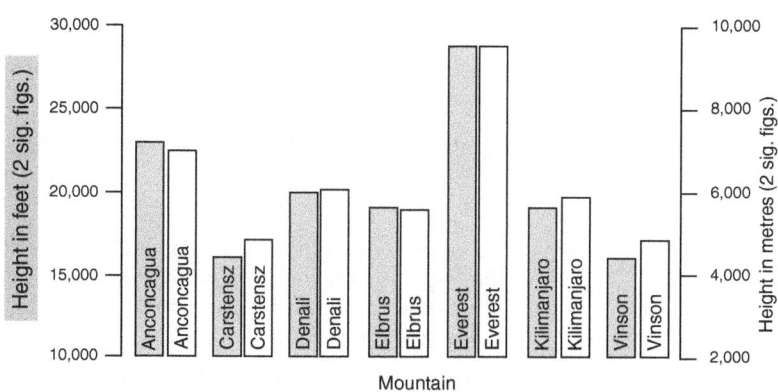

The significance of significant figures

'No measuring device is totally accurate so significant figures let you know how much uncertainty there is.'

Significant figures show how accurate or precise a measurement is. The word 'significant' means important.

Teaching tip

Clarify the difference between accuracy and precision. A measurement is accurate if it is close to the true value. A measurement is precise if it is close to another measurement, which may or may not be accurate.

Bonus idea ★

Numbers in contexts

As a class, consider the number 675. Suggest scenarios that require different levels of accuracy:

- A pizza costs £6.75. (3 sig. figs.)
- Wood is sold by the metre. I need 675 cm of wood. I need to buy 7 m. (1 sig. fig.)
- The mean temperature in May was 68° Fahrenheit. (2 sig. fig.)

Students then consider four-digit numbers in contexts that require different levels of precision. They can use dice to generate four-digit numbers.

Too often students see work with significant figures as an exercise in rounding numbers in order to simplify them. The question to ask is: 'How accurate does the number have to be?' In science and engineering, every digit in a number might be essential.

Discuss which digits in a number are significant and draw up rules to define them. Then, consider numbers in contexts to find the degree of accuracy required.

Define significant figures

Students work in pairs to discuss the value of the digits in each number, the special case of zero, and the role of the decimal point.

Ask: 'How many significant figures are there in each of these numbers?'

235 6054 5.329 4.06 4.060 0.375

Students write a short explanation of their findings. The teacher summarises the ideas on the board. Students discuss them to find the rules governing significant figures:

- Digits from 1 to 9 are significant in a number.
- Zero is significant when it is a placeholder.
- The final zero in a decimal is significant. It is called the trailing zero.
- A single zero before the decimal point is not significant as it merely draws attention to the decimal point.

Roots, powers and standard form

Part 5

Rooting out square roots

'Working with roots of numbers has applications in many areas of maths, including geometry, trigonometry and graphs.'

Explore models of square numbers to find the square root. Finding the square root of a number is the inverse operation to squaring a number.

The square root of a number is the quantity that is multiplied by itself to give a squared number. A square root can be positive or negative.

Square roots

- Students work in two groups to construct four large models of squares using base ten material.
- They then exchange places and work out the square root of each other's models. They should find two different ways of finding the root: either counting or measuring the length of one side. They then calculate the number of items in the square.
- Students summarise the calculation in their own words and record the results in a table. They use N to represent any number.
- They then model all the square numbers up to 100.

Example: Length of one side of the square is 7 so the number of items in the square is 7 x 7 = 49. Therefore $7^2 = 49$ so $\sqrt{49} = 7$.

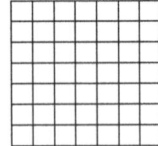

Length of one side of square	Number of items in model N^2	Square root $\sqrt{N^2}$
7	$7^2 = 49$	$\sqrt{49} = 7$

Squares in a row

A quick game for two to four players. Each player draws a table with two rows. The top row is called N and contains the numbers 1 to 10. The second row is called N^2. Players fill this in during the game.

- Players take turns to roll a 1–10 dice.
- They calculate the square of the number and write it in their own table.
- If a square is already filled, play passes to the next player.
- The winner is the first player to calculate three adjacent numbers on their own table.

N	1	2	3	4	5	6	7	8	9	10
N^2										

More rooting around

'This method makes it so much easier to find a sensible number to start working with to find the square root.'

Help students to find the approximate square root of numbers that are not perfect squares through reasoning using number lines and diagrams.

Show students how to reason from perfect squares to find an approximate square root, then use trial and error to refine the search. Using a number line and area diagram helps to identify the best starting point. Visualising the relationship between a number and the nearest square numbers helps them understand the calculation. Transferring the proportions from a number line to an area diagram helps identify the starting point for the calculation.

As a class, estimate the square root of 23. Students should each do their own calculations on squared paper.

- Draw an empty line and mark the midpoint, 23 and the two square numbers that 23 lies between. 23 is about three quarters of the distance from 16 to 25.

- Use the same proportions to show the value and positions of the square roots on a second number line.

- Draw a diagram showing that $5^2 = 25$ squares.

- Now use a dotted line to draw the approximate position of 23 as a square. This is shown as a dotted line on the diagram. The ratio of the lengths of the sides of the square is proportional to the distances between the numbers on the number line. So the length of one side of the new square is approximately 4.75.

Taking it further

Square roots are important in many areas of life that involve measurement. Ask students to list ten jobs that use square roots.

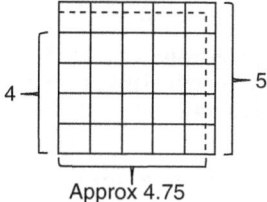

4 —

— 5

Approx 4.75

- Check the accuracy of the approximation 4.75^2 and adjust it until you find a reasonable estimate. It is only necessary to do calculations to two decimal points:

4.75 x 4.75 = 22.56

Answer slightly too low. Try 4.80.

4.80 x 4.80 = 23.04

23.04 rounds to 23 so this is a good estimate for the square root of 23.

Positive and negative square roots

'A graph accentuates the positive result!'

A square number is the result of multiplying a whole number by itself. Since a negative multiplied by a negative will always result in a positive answer, the square number will always be positive.

Students often do not understand why the square root can be either positive or negative. They may not realise that a negative sign attached to a number is different to the minus operator, which means subtract. Try modelling the following activities with students to help them grasp this concept.

Square numbers

- Square the following numbers, show the working and record the results in a table: 4, −4.
- Attach a plus sign to the front of the positive numbers to emphasise the importance of the sign. Use brackets to make it clear that the sign is part of the number.

Number	Number squared	Workings
4	$(+4)^2$	$(+4) \times (+4) = 16$
−4	$(−4)^2$	$(−4) \times (−4) = 16$

Square roots

- Find the square roots of the following numbers and record them in a table: 36, 64.

Number	Square roots
36	$\sqrt{36} = \pm6$
64	$\sqrt{64} = \pm8$

- Write a brief explanation of the results:

Example: 6 x 6 = 36 so √36 = 6
−6 x −6 = 36 so √36 = −6
Therefore, the square root of 36 can be positive or negative: √36 = ±6

Squares and square roots

- Plot squares and square roots for the numbers 1 to 6 on graph paper.
- Pay careful attention to the scales used.
- Draw up a table to show the x and y values.

x	1	2	3	4	5	6
y	1	4	9	16	25	36

- Join the points to draw the graph. Note that this graph will be curved.

Taking it further

Students learn the rule that a negative number cannot have a square root. However, the imaginary number 'i' is defined as the square root of minus one (√−1). Ask students to explore the importance of 'i' in science and engineering.

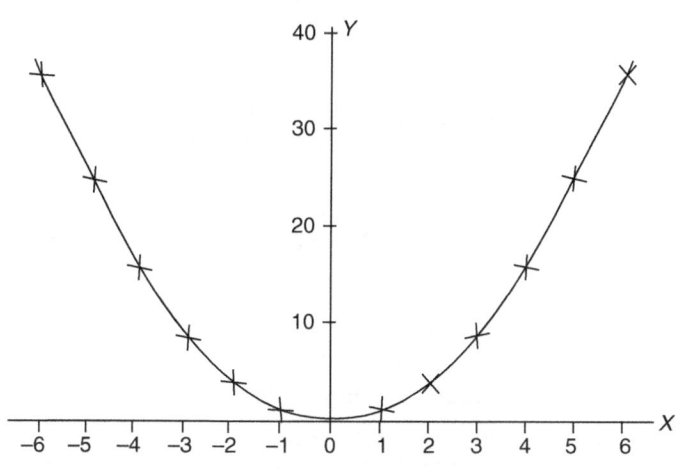

Graph showing the squares and square roots of numbers to 6

Power in the chain

'The idea of multiplying a number by itself may seem pointless; actually it shapes our lives. Powers appear in many subjects, including geography, physics, economics, chemistry and computing.'

Power, index and exponent all mean the same thing. They show how many times a number is multiplied by itself.

Teaching tip

It quickly becomes impossible to model the results of the equations. Stop the clock and discuss this when the problem arises. This is exponential growth in action (see Idea 48).

Taking it further

To take the 'exponential growth in action' activity further, ask: 'How big do you think the quantity will grow to after another two minutes?' Each pair makes a prediction and the teacher writes them on the board. The students continue modelling for two more minutes before stopping and discussing the findings.

Index numbers (also powers or exponents) indicate how many times a number is multiplied by itself. This is called exponential growth. Powers were introduced with square numbers, for example: $4^2 = 4 \times 4$ (see Idea 18). In 4^2 the base number is 4 and 2 is the power.

Power grab

Students play in pairs and take turns to write the chain of a number multiplied by itself and then write it in power notation:

- Roll a 1–6 dice to generate a base number.
- Roll a 1–10 dice to generate the power or index.
- Write the result in the expanded form and then write it using power notation.
- The largest numbers wins.

Exponential growth in action

Students work in pairs to investigate exponential growth. Do not tell the students what to expect: the quantity will grow slowly initially then the growth will accelerate rapidly. Set the task: 'Model how many times you can multiply 2 by itself in two minutes.'

- Start with two objects and model the result of each stage in the multiplication chain.
- Draw up a table to record all the results.
- Write the equation in expanded form.

Discuss the results from each team as a class.

Multiplying and dividing powers

'This quick method only works if the quantities in the calculation have the same base.'

Multiply powers by adding the index numbers; divide powers by subtracting the index numbers. Use cards to demonstrate why the rule works.

Multiplying powers

- Use cards to expand the question and show what the power notation means, e.g.:

$3^5 \times 3^2 =$ | 3 | ×3 | ×3 | ×3 | ×3 | × | 3 | ×3 |

- Write the expansion. Using brackets helps to keep track of the original elements. Explain that you are counting how many times the number 3 is written, then writing the answer as a power of 3 to show how many threes there were: $(3 \times 3 \times 3 \times 3 \times 3) \times (3 \times 3) = 3^7$.
- Ask, 'Is the result the same if you add the index numbers 5 and 2?' Write: $5 + 2 = 7$.
- Compare the results with the powers in the equation: $3^5 \times 3^2 = 3^7$.
- They are the same so the rule works.

Dividing powers

- Prove that to divide powers you subtract the second index from the first.
- Model the division and write it as a fraction. Simplify the fraction by cancelling. Write the answer as a power:
- Ask, 'Is the result the same as subtracting the power 2 from the power 5?' Write: $5 - 2 = 3$.
- Compare the results with the powers in the equation: $3^5 \div 3^2 = 3^3$.
- They are the same so the rule works.

$3^5 \div 3^2$

$= \dfrac{3 \times 3 \times 3 \times \cancel{3} \times \cancel{3}}{\cancel{3} \times \cancel{3}}$

which is 3^3.

The power of geometric growth

'An index looks like a tiny number. Now I see why they have such enormous power.'

Compare linear sequences and geometric sequences to understand the powerful force of exponential growth. Harnessing the power of exponential growth can change lives — for good or bad — so show students what it looks like.

Teaching tip

Ask students to draw a diagram of the model and use a different colour to show the value of each new term.

In a linear sequence, you start with a quantity and add a fixed amount to find each term. This is the model for multiplication as repeated addition. The quantity increases at a steady rate. In a geometric sequence, you start with a quantity and multiply it by a fixed amount. Then, each new term is a multiplication of the previous amount in the sequence. Initially, the value of the quantity gets bigger fairly slowly but the speed of growth quickly increases.

Students work in pairs to discuss, model and record the sequences for the quantity 3. At this stage the number of each term is omitted as it can cause visual confusion in the table. However, students who wish to do so can include it.

Linear sequence		Geometric sequence		
3 x 1	3	3^1	3	3
3 x 2	6	3^2	3 x 3	9
3 x 3	9	3^3	3 x 3 x 3	27
3 x 4	12	3^4	3 x 3 x 3 x 3	81
3 x 5	15	3^5	3 x 3 x 3 x 3 x 3	243

Example: Modelling the 5th term in both sequences

Linear sequence: 3 x 5 = 15

The quantity 3 is repeated 5 times:

Geometric sequence: 3^5 = 243

Each quantity is 3 times the result in the previous term:

Introducing standard form

'I get very confused with the jumping decimal point so looking at the place value grid works better for me.'

Standard form makes very big or very small numbers easy to write and to compare. A number in standard form is written as a quantity between 1 and 10 multiplied by a power of 10. This can be summarised as $A \times 10^n$ where $1 \leq A > 10$ and the power n is an integer that can be positive or negative.

Teaching tip

Discuss the vocabulary with students. In the UK, the term 'standard form' is used interchangeably with 'scientific notation'. However, the word 'standard' means 'the accepted way'. In maths, standard form is also used in other contexts. It may refer to the expanded form of a number ('partitioning' in the UK), or the way of writing a linear equation as 'expression' = 0.

Students need a secure concept of the place value system and decimal numbers. Refresh their knowledge of the place value grid and link it to standard form. The rule for finding the index number in standard form is often taught as how many times the decimal point moves. Some students find this explanation confusing. Try this method, which focuses on the value of the number in the place value system:

- Record a number on a place value (PV) grid and multiply it by 10 to show that each PV position differs by a power of 10.
- Then, write each number in standard form as a number greater than 1 and less than 10 multiplied by a power of 10.
- The index number in the power of 10 states how many times it is multiplied by 10 as shown in the place value positions. The value of 10^0 is 1. Below, $6 \times 10^0 = 6 \times 1$, which is 6.
- Students discuss the pattern and write their own explanation of what has occurred.

	M Th HTU HTU HTU	Standard form
6×1	6	6×10^0
6×10	60	6×10^1
$6 \times 10 \times 10$	600	6×10^2
$6 \times 10 \times 10 \times 10$	6 000	6×10^3
$6 \times 10 \times 10 \times 10 \times 10 \times 10 \times 10$	6 000 000	6×10^6

Now, students work in pairs to practise applying the rule to multi-digit numbers:

- Ask them to investigate 35 and place it on the place value grid. Students need to discover for themselves that now there will be a decimal point and numbers after the decimal point.
- Discuss the results as a class.
- Students then express the following numbers in standard form:

536   3,759   673,827

- Finally, students write the following numbers as conventional numbers and then in standard form:

five billion [Answer: $5{,}000{,}000{,}000 = 5 \times 10^9$]

forty one thousand and fifty six [Answer: $41{,}056 = 4.1056 \times 10^4$]

Miniscule numbers

Discuss with the students how you might represent numbers that are smaller than 1. Here it helps to use the concept of a fraction as one number divided by another number. Consider the fraction $\frac{6}{10}$. This can also be expressed as $6 \div 10$ or $6 \times \frac{1}{10}$. If students are unsure, revise the link between fractions and decimals (Idea 30).

Record the numbers on the place value grid and write the standard form as a negative power. First, show the results in the form of n multiplied by $\frac{1}{10}$ then as n divided by 10 to make it clear that the meaning is the same.

	H T U$._{\frac{1}{10}}$ $_{\frac{1}{100}}$ $_{\frac{1}{1000}}$	Standard form		H T U$._{\frac{1}{10}}$ $_{\frac{1}{100}}$ $_{\frac{1}{1000}}$	Standard form
6×1	6	6×10^0	$6 \div 1$	6	6×10^0
$6 \times \frac{1}{10}$	0.6	6×10^{-1}	$6 \div 10$	0.6	6×10^{-1}
$6 \times \frac{1}{10} \times \frac{1}{10}$	0.0 6	6×10^{-2}	$6 \div 100$	0.0 6	6×10^{-2}
$6 \times \frac{1}{10} \times \frac{1}{10} \times \frac{1}{10}$	0.0 0 6	6×10^{-3}	$6 \div 1000$	0.0 0 6	6×10^{-3}

Students do examples. They write the standard form and show it on the place value grid.

Display a list of interesting numbers in standard form on the classroom wall. Ask each student to find one number and briefly explain why it interested them. Examples: What is about 1.5×10^8 km away? The Sun. What measures approximately 1×10^{-5} nanometers? The thickness of a human hair.

Bonus idea ★

See the online resources for a worksheet that can be used to practise standard form.

Multiplying and dividing with standard form

'Scientific notation is much easier than writing out all those zeros.'

Develop students' confidence in multiplying and dividing numbers in standard form. This is important as it is used extensively in science and engineering.

Taking it further

Consider real-life calculations using standard form. Students show their working and ensure they work in the correct units of measurement (see Idea 84). For example: Earth is 150 million km from the Sun. The speed of light is 3×10^8 metres per second. How long does it take a beam of light to reach Earth?

Answer: Time taken is distance divided by speed (see Idea 93). Distance = 1.5×10^8 km. Speed = 3×10^5 km per second. So:

$$\frac{1.5 \times 10^8}{3 \times 10^5} = \frac{1.5}{3} \times \frac{10^8}{10^5}$$

$= 0.5 \times 10^3$ seconds

$= 5 \times 10^2$ seconds

It takes about 500 seconds for a beam of light to reach Earth from the Sun.

Rather than teach a set of rules, help students understand the calculations. Students have learned how to multiply and divide numbers with exponents (see Idea 47). Now, they apply this knowledge to calculating with standard form. Make sure that students understand that standard form in this context is also called 'scientific notation'. The term 'standard form' is used in a different sense in algebra.

Students work in pairs to discuss and solve the following questions with answers given in standard form. They need to sort out what they will call the different numbers in the standard form expression if they are to avoid confusing themselves and each other. For clarity call the number in the front A and use n to denote the power of 10: $A \times 10^n$.

$(8.2 \times 10^3) \times (0.5 \times 10^2)$
$(3.68 \times 10^7) \times (1.1 \times 10^4)$
$(5.42 \times 10^7) \times (3.9 \times 10^4)$
$(7.4 \times 10^4) \div (0.5 \times 10^2)$
$(2.75 \times 10^5) \div (1.1 \times 10^{-4})$
$(3.8 \times 10^{-3}) \div (7.6 \times 10^{-5})$

Discuss the results as a class. Consider the following questions:

- Do you need to rearrange the terms in the calculation?
- Do you need to adjust the product of the numerical part of the calculation?

Ratio and proportion

Part 6

Ratio and proportion

'Make sure students can work with basic ratios involving two quantities before embarking on ratios with more than two parts.'

Ratios show the relative size of one quantity compared to another. A part-to-part ratio describes the relationship between two distinct groups. A part-to-whole ratio shows the relationships between a sub-group and the whole group.

Ratios are important for everyday problem solving, from cookery to currency exchange to reading maps. Use objects to model the concept of ratios, then ask students to pose questions involving data.

Part-to-part ratio

Use different-coloured counters to demonstrate ratio. Students work in small groups. Each student models and records their own ratio. They discuss the results with the group.

- Ask students, 'What is the ratio between black and white counters?'
- The students take a handful of counters and separate them into two groups according to colour.
- They draw a diagram to show the number of objects in each group.
- They express the relationship as a ratio in the form of two numbers separated by a colon.
- Finally, they devise a word problem using the ratio and describe the ratio to the other students.

Example: ●●● ○○○○

There are 3 black counters and 4 white counters. The ratio of black to white counters is 3:4.

There were 3 girls and 4 boys in each basketball team. How many girls were playing in the match?

Part-to-whole ratio

Build on the concept of splitting numbers into their components (see Idea 11).

- Pose the question: 'What is the ratio between black counters and the total number of counters?'
- Now, students draw a triad diagram to make the relationship between a whole quantity and its components clear.

Example:

There are 7 counters. 3 counters are black. The ratio of black counters to total counters is 3:7.

Taking it further

Compare more than two parts. Use horizontal brackets to depict ratios involving more than two parts. In this example, students find the ratio between squares, circles and triangles by sorting them and drawing a pictogram. The horizontal brackets and the numbers below make the ratio of 4:2:3 clear.

Cooking it up

'Baking biscuits is a great way to make ratio and proportion popular.'

Help students learn how to change a quantity according to a ratio. From mixing concrete to baking cakes, ratios are used all the time to work out how much of each ingredient is required to make large-scale quantities.

Work through this example of changing quantities for baking. Then, discuss other situations in which quantities have to be changed and ask students to frame their own questions that require scaling quantities up or down.

Explain that you are going to practise applying ratio to find the amount of each ingredient required to bake 3 kg of shortbread biscuits:

- The ingredients for making a small quantity of shortbread are butter (125 g), sugar (75 g) and flour (175 g).
- State the ratio of butter to sugar to flour and simplify. (125:75:175 simplifies to 5:3:7)
- Add all the ratios to find the total number of parts involved. (5 + 3 + 7 = 15)
- Express each ingredient as a fraction.
- Find the value of each fraction of 3 kg to calculate the quantity of each ingredient.

Establish the habit of putting information into tables. Draw up a table to show the ratios, each ratio as a fraction, and the quantity of each ingredient needed to make 3 kg of shortbread.

	Ratio	Ratio as a fraction	Quantity
Butter	5	5/15	5/15 x 3 kg = 1 kg
Sugar	3	3/15	3/15 x 3 kg = 0.6 kg
Flour	7	7/15	7/15 x 3 kg = 1.4 kg

Plan to scale it down

'I can now relate plans and diagrams to real life.'

Ratio is the basis of using scales for measurement such as for road maps or plans of buildings. Investigate ratio as scaling by measuring and drawing a plan to scale.

Students work in pairs and use a measuring tape to measure the dimensions of the classroom in metres and then scale them down to draw a plan:

Teaching tip

Working with ratios and scales lays foundations for working with graphs.

- The students discuss a suitable ratio scale. (A commonly used scale is 1 cm on the plan to represent 1 m (100 cm) on the ground, giving a scale factor of 1:100.)
- They use a measuring tape to measure the length of each wall, the width of windows and doors, and their position on the walls. They record the measurements on a sketch.
- They show the actual measurement and the scale measurement in a conversion table.

Taking it further

Encourage students to plan and map larger areas such as school playing fields or a local park. This activity provides valuable practice in converting between units of measurement and developing a sense of the relative size of those units.

	Actual measurement (cm)	Conversion factor	Plan measurement
Wall A window	60 cm 240 cm 230 cm	1/100	0.6 cm 2.4 cm 2.3 cm
Wall B	400 cm	1/100	4 cm
Wall C door	330 cm 90 cm 110 cm	1/100	3.3 cm 0.9 cm 1.1 cm
Wall D	400 cm	1/100	4 cm

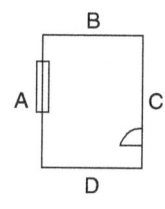

- They draw an accurate scale drawing of the classroom using a scale factor of 1:100. Make sure they include a scale bar to show how long 1 unit of measurement is on the plan (see online resources for an example).

Map reading

'Applying ratios ensures that a scale drawing is the same shape as the original but a different size.'

This activity will help students interpret scale drawings.

Students use a map to plan a cycle ride and calculate the actual distance of the ride using the scale on the map.

- Plan the route and draw a line to mark it on the map. Discuss which landmarks to use to identify the measurements taken (e.g. villages or railway crossings). Mark them on the map.
- Measure the distance on the map in centimetres (cm). Record each measurement and list them in a table.
- Find the total length of the ride in cm.
- Use the scale on the map to calculate the actual distance. Tabulate the information to show the relationship between the scalar lengths and the actual distance (see Idea 56).
- Do the calculation and show all the steps of working. Convert the result into appropriate units of measurement.

	Distance on map	Actual distance
Planned ride	14 cm	D cm
Scale	1 cm	25,000 cm

Answer: If 1 cm represents 25,000 cm, then 14 cm represents 14 x 25,000 cm.
1 km is 100,000 cm.

So, $D = 14 \times 25,000$ cm
* $= 350,000$ cm*
* $= 350,000 \div 100,000$ km*
* $= 3.5$ km*
The cycle ride will be 3.5 km long.

Human scale

'Scale changes your view of the world around you.'

Explore with students a few ways in which size, scale and proportion affect the way that we interpret what we see.

The human face

Ask students to draw emojis of faces for a baby, a young adult and an elderly person. They should use only a circle or an oval, eyes and mouth, e.g.:

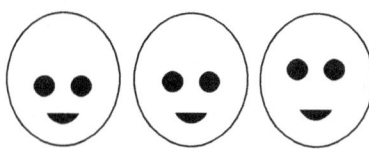

Students work in small groups to discuss their drawings and how successful they were. The key feature in drawing a human face is the position and proportion of the eyes in relation to the head. Imagine a horizontal line in the middle of the face: a baby's eyes are slightly below the line. The eyes are large in proportion to the mouth and they are close together. As a person grows, the bones of the face get bigger so the eyes appear about halfway up the face and are further apart. In old age, the eyes are above the midline.

Human scale

Display a series of numbered photographs of different buildings in the following categories: living, working, worship. For example, living spaces might include a cottage, high-rise flats, a prison and a tent. Ask students to write the feelings that each one evokes in them. Write the responses to each one on the board and discuss how the proportions and the scale of the building generate particular emotions.

Teaching tip

Human scale is the relationship between something and the size of a human being. Understanding scale plays an important part in interpreting graphs and other visual data representations.

Taking it further

Cartoonists Gerald Scarfe and Peter Brookes aim to evoke a particular response from the viewer by distorting the proportions of faces and using unusual scales to create the effect. Consider why this approach is so effective.

Universal conversion grid

'A great security blanket for students who forget the rules, as well as helping them to focus on what the question is asking.'

Help students do conversions easily and with confidence by identifying what you want to know and what you do know, and applying knowledge of proportions to find the conversion factor.

Arrange the information in a simple grid to show the relationships between the quantities. In the early stages, use a question mark (?) to denote the unknown and use the word 'what?' to draw attention to it. Later, this can be replaced by x. The advantage of this method is that it requires students to analyse the information in the question in order to decide what is relevant and to choose appropriate labels for each column and row. Effective analysis is key to problem solving.

First, complete an example with the whole class: Ben cycled the 874 miles from Land's End to John O'Groats in 300 hours. How many days did it take him?

As always, discussion is essential. Ask students to suggest questions to extract the relevant information:

- What do I want to find out? The time taken in days.
- What information do I have? The ride took 300 hours. The conversion factor is 1 day = 24 hours. The length of the journey is not relevant in this case.

- Can I create a diagram to make the problem clearer?

	Time taken	Conversion factor
Days	?	1
Hours	300	24

- Can the calculation be expressed as an equation between two equivalent fractions?

$$\frac{?}{300} = \frac{1}{24}$$

$$\frac{300 \times ?}{300} = \frac{1}{24} \times 300$$

Isolate the unknown by multiplying both sides by 300 and cancelling.

$? = 13$

Answer: The ride took 13 days.

- Does the result make sense?

Students now work in pairs to solve the following problems. Each student has a turn giving instructions that the other person follows. For the next question their roles are reversed. If the instructions are unclear, they ask for clarification.

- My exam score was 28 out of a possible 40 marks. What percentage did I score?
- The train travels at 70 miles an hour. How long will it take to travel 35 miles?

Taking it further

This grid applies to any conversion task: units of measurement, speed calculations, ratios of sides of geometric figures, pie charts, graphs, currency conversions, scaling on maps or building plans, percentages – anywhere that one quantity is changed in fixed proportion.

Rates of change: percentage increase

'Understanding percentage increase has helped me compare value.'

Tabulate information to keep it under control. This method also makes it clear why the final percentage amount is the multiplier.

Make sure that students are confident working out equivalent fractions before embarking on percentage calculations. Percentages can easily be expressed as a decimal, which is essential for calculations using a calculator.

The steps in the calculation are as follows:

$$\frac{?}{25} = \frac{120}{100}$$

$$\frac{?}{25} \times 25 = \frac{120}{100} \times 25$$

$$? = \frac{120}{4}$$

$$? = 30$$

Answer: The actual cost of the jeans is £30.

Work through this problem with the students:

The basic price of a pair of jeans is £25 without VAT. Find the actual cost of the jeans including VAT at 20%.

- Discuss what questions need to be asked to analyse the information. Write the questions on the board. This seems simple to those who already understand but it is at this stage that students with difficulties struggle.
  - What do I want to find out? [The actual cost of jeans]
  - What do I know? [Price excluding VAT = £25. VAT = 20%.]
  - Can I express the actual cost to the basic price as a ratio? $[\frac{?}{25}]$
  - What does 100% represent in this example? [The basic price excluding VAT]
  - What is the new cost as a percentage of the basic cost? [100% + 20% = 120%]
- Record the information in a table.
- Write the equation to show the price and the percentage as ratios. Solve the equation and show all the steps in the calculation.

	Price (£)	Percentage (%)
Actual cost	?	120
Basic cost	25	100

The rule of 72

'So many teenagers get into debt because they don't realise the effect of compound growth.'

The rule of 72 is a quick way to compare the costs of borrowing money at different interest rates.

Many teenagers incur debts that rapidly become hard to repay. It is difficult to understand the power of compound growth by considering the interest rate in the first few years. Amounts that grow exponentially start growing gradually and then increase dramatically. Show the staggering effect of compound growth rates by using the 'rule of 72' to find the doubling time for an amount. Doubling time is the length of time it takes for a quantity to double in size at a constant rate. Find the number of years by dividing 72 by the percentage growth rate. The rule is derived from the formula $R \times T = 72$ where R is the annual growth rate and T the doubling time.

Students work in pairs to investigate the effect of the rate of interest on borrowing. The examples below are representative of different kinds of debt – a bank loan, credit card debt and a payday loan.

Question: You borrow £100. If you pay nothing back, how long will it take the debt to double...

- ... if the annual interest rate is 6%?
- ... if the annual interest rate is 24%?
- ... if the annual interest rate is 1,200%?

Use the formula $T_{years} \approx 72/R$ and record the results in a table. Discuss the results with the students and ask them which organisations might charge the different rates. The answer table is provided in the online resources.

Teaching tip

Use the rule of 72 for growth rates up to about 20%. It is less accurate for higher rates of growth where the rule of 70 gives better estimates.

Taking it further

Explore the exponential growth of human populations and their effect on the environment. As human populations increase, they impact wildlife populations, which decrease.

The sale price: percentage decrease

'Understanding percentage decrease has helped me find the best deal.'

The key to success with percentage decrease is analysing the information. Percentage decreases are slightly trickier as you have to be clear whether 100% represents the original amount or the decreased amount.

Teaching tip

Encourage students to quickly sketch number lines to help them compare the information (see Idea 31).

Follow the same procedure as Idea 57. Analyse the information with students by questioning and then summarise it in a table. Practise both slightly different approaches to encourage students to be flexible in their thinking. In the first example, the original price is known, and in the second case the sale price is known. Start with simple examples to build confidence and gradually introduce more complex situations. Encourage students to write down pertinent questions and summarise information themselves.

Use this problem as a starting point:

A bike is on sale for 15% off the original price of £600. What is the purchase price and how much is the saving?

- What do I want to find out? [The sale price]
- What do I know? [Original price = £600. Discount = 15%.]
- What does 100% represent in this example? [Original cost]
- What is the sale cost as a percentage of the original cost? [100% − 15% = 85%]

	Price (£)	Percentage (%)
Sale cost	?	85
Original cost	600	100

$\frac{?}{600} = \frac{85}{100}$

$\frac{?}{600} \times 600 = \frac{85}{100} \times 600$

$? = 510$

Answer: The sale price of the bike is £510.

Now try this example with the students:

A bike is on sale for 15% off the original price. The sale price is £510. What was the original price?

- What do I want to find out? [The original price]
- What do I know? [Sale price = £510. Discount = 15%.]
- What does 100% represent in this example? [Original cost]
- What is the sale cost as a percentage of the original cost? [100% − 15% = 85%]

	Price (£)	Percentage (%)
Original cost	?	100
Sale cost	510	85

$\frac{?}{510} = \frac{100}{85}$

$\frac{?}{510} \times 510 = \frac{100}{85} \times 510$

$? = 600$

Answer: The original price of the bike was £600.

Money surprises

'When I look at percentages, I panic.'

Managing money is an essential life skill and percentages play a key role. Grab students' attention by exploring the effect of interest rates on borrowing and saving.

There are two kinds of interest: simple and compound. Work through an example of each, step-by-step with students to help them understand the important difference between the two.

Interest is the amount it costs to borrow money. It is expressed as a percentage of the principal, that is the initial amount borrowed. Interest can be positive, such as money earned on savings, or negative, as in the debt owed on borrowed money.

- Students spend two minutes jotting down what they know about the term 'interest' and words associated with it.
- Summarise the main ideas on the board.
- Students can suggest examples of interest payments from their own experience.

Now, complete worked examples with the students for simple interest and compound interest. Put the information in tables and work out an equation for the calculation.

Simple interest

Here the same amount of interest is paid each year. The price of a car is £4,000. A loan is available at 5% over 3 years. How much do you pay for the car?

	Principal amount	5% interest	Total amount owed at end of year
Year 1	£4,000	200	4,200
Year 2	£4,000	200	4,400
Year 3	£4,000	200	4,600

The total cost of the car is
£4,000 + (£4,000 x 0.05 x 3) = £4,600

Compound interest

Here the interest rate stays the same but a different amount is paid each year. Interest is paid on the principal amount plus the interest from the previous year. You save £4,000 at a rate of 5% over 3 years. How much do you have after 3 years?

	Principal amount	5% interest	Total amount saved at end of year
Year 1	£4,000	200	4,200
Year 2	£4,200	210	4,410
Year 3	£4,410	220.50	4,630.50

The total amount saved is
£4,000 (1.05)^3 = £4,630.50

Taking it further

A difference of £30.50 may seem quite small. Compound interest is exponential growth so the amount of interest payable increases dramatically as the length of time increases (see Idea 100).

Algebra

Part 7

Why collect terms?

'Algebra is easy when I do it this way.'

Algebra is incomprehensible if you present students with a series of numbers and letters and tell them to learn the rules. Colour and counting are all you need to give students a memorable start to algebra.

The equipment is simple: counters in four different colours. (Maths classrooms often have a large box of coloured counters, otherwise a pair of students will need a minimum of 20 counters in four colours to work with.)

- Ask students to sort the counters into colours. Each student collects their own piles. They count the number of counters of each colour and write a sentence to summarise their collection. Use the word 'and' instead of a comma as this helps emphasise the separateness of each group. E.g. 'I have 4 red and 3 green and 6 yellow and 3 blue counters.'
- Next, ask students to summarise the statement by using the initial letter of each colour word and an addition sign instead of the word 'and'. The letter in each term is called the variable. E.g.: $4r + 3g + 6y + 3b$
- This is an algebraic expression consisting of four terms. Now students work in pairs to combine their expressions. Both students write down the two expressions. They do not move the counters. E.g.:
$4r + 3g + 6y + 3b + 6b + 2g + 3r + 7y$
- Each student collects like terms according to the variables. It is conventional to put variables in alphabetical order. E.g.:
$3b + 6b + 3g + 2g + 4r + 3r + 7y + 6y$
- They then simplify by adding like terms:
$9b + 5g + 7r + 13y$

Algebra: substitution

'If footballers can do it so can maths.'

In substitution you put in an amount for the unknown. This can be seen as the heart of algebra. Solving a given problem from a known formula involves substitution.

Discuss with students how to use known everyday formulae. For example:

How long to cook a chicken

The formula is **cooking time = 15 + mass in grams/500 x 25**. This means the chicken should be cooked for 25 minutes for every 500 grams plus 15 minutes. Ask students to find out how long a 1.1 kg chicken will take to cook. (Make sure they follow BODMAS rules and divide the mass of the chicken by 500 first.)

How much your mobile calls will cost on a particular tariff

The formula for the example tariff is 40 pence per minute after the first free 300 minutes, which can be represented by **cost = (total number of minutes used − 300) x 0.4**. Students may like to investigate how many minutes they use. They can put this into the equation.

Exchange rates

The formula is **number of euros = number of pounds x current exchange rate in euros**. Provide today's exchange rate and ask students to work out how many euros you would receive for £100.

Taking it further

Link with science formulae, including speed, distance, time and density, mass, volume.

Expanding brackets

'I thought they just got wider and wider.'

Brackets are used in maths expressions to avoid ambiguity. However, students often find the brackets confusing when they do calculations.

Students often find 'expanding' or 'multiplying out' brackets difficult in algebra so practise it with numbers first. Build on the area model of multiplication to make the process transparent and to develop proficiency.

First, revise multiplication with numbers with the students:

- Partition numbers into tens and units and rewrite the equation. Note that the 'x' sign is not used between brackets as this can be mistaken for the letter x, which is used for an unknown.

$13 \times 24 = (10 + 3)(20 + 4)$

- Draw the box diagram with the partitioned numbers. There are now four areas. The 'x' at the top left means 'multiply'.

$$
\begin{array}{c|c|c}
\times & 20 + & 4 \\
\hline
10 & & \\
3 & & \\
\end{array}
$$

Write and solve the equation for the area of each rectangle. Then, write the products in the relevant box.

$$
\begin{array}{c|c|c}
\times & 20 & + \; 4 \\
\hline
10 & 200 & 40 \\
\hline
3 & 60 & 12 \\
\end{array}
$$

$10 \times 20 = 200$
$10 \times 4 = 40$
$3 \times 20 = 60$
$3 \times 4 = 12$

- Add all the results to find the total value of 13 x 24.
- Describe the model using equations and solve:

$(10 \times 20) + (10 \times 4) + (3 \times 20) + (3 \times 4)$

$= 200 + 40 + 60 + 12$

$= 312$

Now replace numbers with letters:

- Explain that you will be expanding the brackets for $(a + b)(c + d)$.
- Draw the box diagram and write the products in each rectangle.
- Write and solve the equation for the area of each rectangle. Note that the multiplication sign is not used between letters. Simply writing the letters next to each other indicates multiplication.
- Then, write the products in the relevant box:

×	c	d
a	ac	ad
b	bc	bd

a multiplied by $c = ac$
a multiplied by $d = ad$
b multiplied by $c = bc$
b multiplied by $d = bd$

- Add all the results to find the total value of the expanded brackets:

$(a + b)(c + d) = ac + ad + bc + bd$

Algebra: the beginnings

'So, what is algebra for?'

When algebra is first broached in the maths classroom, fear can be seen in many faces. This exercise together with discussion helps students realise that algebra is all around us and is accessible to all.

Divide students into two groups. Students will move around to each activity. (If you have a larger class, divide them into six groups and double up so three groups are doing each activity simultaneously.)

Activity 1

- Lay out on a table four cards, on three of which you have written numbers picked from 1 to 10. The fourth card is blank.
- Tell the students that the total of the cards is 20.

$$\boxed{2}\ \boxed{7}\ \boxed{5}\ \boxed{}\ \boxed{=}\ \boxed{20}$$

- Ask the students to find the value of the blank card.
- Ask the students to write out the sum: $2 + 7 + 5 + ? = 20$
- Discuss with the students what they should do to find the answer.
- Point out that they have begun to solve an equation.

Activity 2

- Student A constructs a barrier to hide the pattern.
- Using Cuisenaire rods, create patterns with different colours and placements. The colours could be red, yellow, green and blue. Placements should be vertical and horizontal.

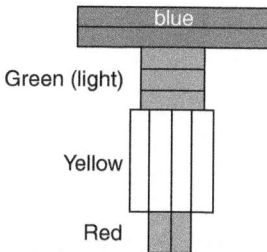

Green (light)

Yellow

Red

Taking it further

Discuss what algebra has in common with the following (use the images in the online resources):

- Musical notes: a visual representation telling us how to perform music.
- Scientific formulae: they involve algebra and instruct us how to solve problems.
- Coding: a language instructing a computer to work.
- Recipes: instructions telling us how to perform a task.

- Student A uses directional language to describe the pattern so that the other students can recreate it.
- Replace the names of rods and their placement with the initial letter of the rod colour, how many there are and their position, e.g. 2*b* horizontal meaning two blue rods placed horizontally.
- Discuss what use this could have in the everyday. Try to elicit the words 'to describe a process' and 'what to do'.

Sign aware

'I understand substitution but now I have to deal with the signs.'

In this activity students will get to grips with the effect of signs on the process of substitution.

Students may be aware of the rules and this aide-memoire:

A **minus** and a **minus** makes a **plus**. The signs are the same so they are **happy**.

A **plus** and a **plus** makes a **plus**. The signs are the same so they are **happy**.

A **plus** and a **minus** makes a **minus**. The signs are different so they are **unhappy**.

A **minus** and a **plus** makes a **minus**. The signs are different so they are **unhappy**.

Explain to students that this is the same for multiplying.

Substitution of positive numbers in a graduated way

Remind students what a number next to a letter means, e.g. 2a means two lots of a. To substitute in the exercise below, students should write in 3 instead of the a and 5 instead of the x. Ask students to work through this exercise:

If $a = 3$ and $x = 5$ what is the value of the following?

- $a + x$ [Answer: 3 + 5 = 8]
- $2a + x$ [Answer: encourage students to write this out: 2 x 3 + 5 = 6 + 5 = 11]
- $2(a + x)$ [Answer: this will become (2 x 3) + (2 x 5) = 6 + 10 = 16]

Substitution with negative numbers

Encourage students to use a number line to complete these calculations:

If **a = -3** and **x = 5** what is the value of the following?

Taking it further

Students should also practise calculating the square of positive *a* and negative *a*, and see what happens (see Idea 45).

- *a* + *x*
  [Answer: -3 + 5 = 2]

- 2*a* + *x*
  [Answer: (2 x -3) + 5 = -6 + 5 = -1]

- 2*a* + 2*x*
  [Answer: (2 x -3) + (2 x 5) = -6 + 10 = 4]

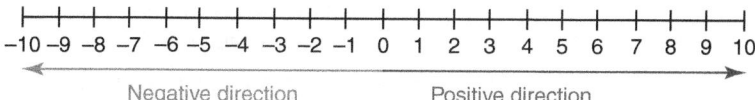

−10 −9 −8 −7 −6 −5 −4 −3 −2 −1 0 1 2 3 4 5 6 7 8 9 10

Negative direction   Positive direction

Variable, coefficient, operation, constant

'It's all too much to remember.'

Too often students are expected to grasp the language of algebra too quickly, which can have consequences for their understanding of the maths.

This activity seeks to establish in the students' minds the definition of each word:

- Using the templates in the online resources, give each pair of students a pack of word cards.
- Students lay these out and discuss the meaning of each term.
- Students write a definition of each term in their own words.
- Now take the expressions provided in the online resources or use ones you have pre-prepared. For example: **$6x + 3a - 7$**
- Students discuss the expression and put out the vocabulary cards to describe each component in the correct order. They cut the expression into individual numbers, letters and symbols, as shown, then put each onto the appropriate vocabulary card.

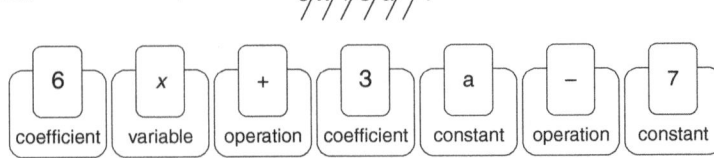

What 'like' is like?

'I am coming to terms with terms.'

This activity is a good lesson warm-up. It gives students practice on what constitutes a like term. It is very important for students to understand this before moving on to multiplication and division of terms. Terms are separated by + or - signs.

Students will identify similarities and differences. This type of activity can aid memory and reinforce understanding. Give students the vocabulary: **variable**, **coefficient**, **operation** (see Idea 66). Then, give them an array of different terms and ask them to identify like terms and explain their answers.

Examples:
7x   x   -2x
Students should be able to say that these are like terms because the variable is x, including the negative term.

$\frac{1}{3}xy^2$ $- 2xy^2$ $6xy^2$ $\frac{xy^2}{2}$

Students should be able to say that the like terms are like terms because the variable is xy^2. The coefficient or sign does not change this.

-3xy   -3   12y² -3x
Students should be able to say that xy, y, y^2 and x are different variables.

Bonus idea ★

Can students identify how many terms there are in the following expressions? Write each expression in correct algebraic form.

$2a - p$
$5 \times z + m \div 2$
$7x - x/a$
$p \div m + qx + r$
$a \times b \times y + z \div 11$

The short hand of algebra

'I now understand that algebra is a language.'

Certain ways of writing algebra can be abstract so it makes it easier if it can be seen as a language. It can also be seen as just a way to shorten the length of a statement.

Taking it further

Ask students: 'What would 4 divided by x look like?' Here students must become accustomed to the fact that one number above another number separated by a line means to divide: $\frac{4}{x}$

Ask students to comment on the statement: 'All languages have rules for the writing of the language as does algebra.' Then, explain that students will have to understand these rules:

1. We do not write 1x or 1y.

Why bother? Writing the letter alone signifies one of them. To explore this further:

- Give students Cuisenaire rods of any colour (e.g. R = red, B = blue, G = green).
- Ask the students to write out what they have in front of them, then to add the items together:
  $R + R + R + G + G + B$
  $G + G + G + B + B + B + B + R$
  $B + B + B + B + B + B + B + R$
- Students should write the answers:
  $3R + 2G + B$
  $3G + 4B + R$
  $7B + R$
- Discuss with students why the single item does not need the number 1 in front of it.

2. A number next to a letter means multiply.

- Ask students to write down x multiplied by 4. They should write: x x 4. Ask them if they see a possible problem with this.
- Ask students to write 8 multiplied by x and they should write: 8 x x.
- This should lead to a discussion of why the 'x' multiplication sign is left out in algebra.
- Explain that the numbers are always written in front of the letters, e.g. $4x$ or $8x$.

So why do we use these letters?

'I thought I was good at maths until letters started to appear.'

In algebra, letters represent a range of unknowns that can be varied, depending on circumstances.

In this activity, students begin to be able to visualise how letters are used to represent the unknown, which is the quantity in the bowl. The students do not know the value of what is in the bowl. Using this activity students begin to see how to build expressions.

- Stars represent what to add or take away from the value or number in the bowl.
- Students can work together. Give students paper and pencils to work with as well as bowls and stars.
- Ask students to imagine a number (*n*) in the bowl.
- Ask students to use the bowl and stars to model, draw diagrams and write expressions for the statements below. The first has been completed as an example.
  - 3 more than the number in the bowl.

$n + 3$

  - 1 more than the number in the bowl.
  - 1 less than the number in the bowl.
  - Twice the number in the bowl and add 2.
  - Subtract 10 from the number the bowl.
  - 4 times the number in the bowl.
  - The number in the bowl multiplied by itself 4 times.

Teaching tip

It is important that students become aware that what is in the bowl can be any amount. If students find this too abstract, begin with a known amount. For example, how do I write:

- 1 more than 45?
- 1 less than 45?
- Twice 45 add 2?
- Subtract 10 from 45?

IDEA 70

Playing with multiplication

'After all that, terms can now combine.'

Students who find the multiplication of terms very confusing may find this exercise helpful.

Taking it further

Ask students to find all the factors of the expressions 12*wv* and then 24*qp*. For example:

12*wv* = (1*w* x 12*v*)
or (2*w* x 6*v*)
or (3*w* x 4*v*)
and (12*w* x 1*v*)
or
(6*w* x 2*v*) or (4*w* x 3*v*).

To begin with, students revise adding simple terms together. Then, explore multiplication to show that they can make a new term out of multiplying two terms together.

Addition

Give students number cards, letter cards and signs from the online resources and ask them to combine them to make these expressions:

$$3w + 4v + w + v$$
$$8q + 3p + 2q + p$$

Students should be able to give a simplified expression for each. They should know that you cannot add terms that are not like terms (where the variable is not exactly the same).

Multiplication

Now give students the following:

$$w \times v =$$
$$q \times p =$$

Using cards, ask them to make new terms: *wv* and *qp*. Discuss what has happened to the multiplication sign. As when using the word 'product' to multiply numbers, multiplication of variables can make a new variable. Finally, give students the first two terms again but this time to be multiplied not added:

- 3*w* x 4*v* (Students should be able to find a number card '12' and multiply *w* and *v* together to make 12*wv*.)
- 8*q* x 3*p* (Students should be able find the number card '24' and combine the *q* and *p* to make 24*qp*.)

What to do with exponents?

'What are these numbers hanging on the letter?'

Students find it difficult to remember to add the exponents when multiplying terms. Understanding what exponents are can aid memory.

Multiplying with exponents

First, discuss with students what 'x squared' means: x is multiplied by itself (see Idea 18). To explore this further, ask students to look at the diagram below:

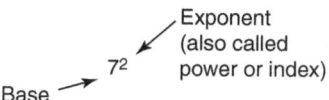

Taking it further

Ask students to look at ab squared. Which letter do you multiply by itself? Explore the difference between ab^2 and $(ab)^2$ by writing them in expanded form. [$ab^2 = a \times b \times b$ and $(ab)^2 = a \times b \times a \times b$]

Now discuss the meaning of the base number or letter. Discuss how 7^2 squared can be written (7×7). Now ask the students to try to write out n^3. They should write: $n \times n \times n$.

So, what do we do when we multiply terms with exponents such as $8n^3 \times 3n^2$?

Ask students to write out the expression: $8 \times n \times n \times n \times 3 \times n \times n$. This is telling us that the base n is being multiplied by itself 5 times so we add the exponents together to arrive at the answer: $24n^5$ (see Idea 47).

Now students can try another example: $4g^3 \times 5g^8$.

Also give them some division examples to work through:

- Simplify n^5 divided by n^2
- Simplify $6w^6y^2 \div 3wy^2$

Baffling brackets

'I am never sure when to use a bracket.'

It is important that students use brackets with ease and can understand why they use them in algebraic expressions.

First, discuss with the students what brackets can do. Try to elicit the phrases 'to make clear', 'separate' and 'plural'. Offer this example:

In a restaurant a man asks for 2 fish and chips. Does he mean 2 fish portions and 2 chip portions or 2 fish portions and 1 chip portion?

Model the different answers and ask the students which one the man meant:

$2(F + C)$
$2F + C$

Now ask students to look at the following statements:

I would like:

- *4 fish and chips.*
- *3 rolls and butter.*
- *5 scones and cream.*
- *6 tables and chairs.*
- *3 bacon and egg.*
- *3 coke and nuts.*
- *2 balls and rope.*

Ask the students to use the initial letter of the items in the statements to write two expressions for each of the above using brackets.

Bracket practice

'Now I see the point of them.'

This activity gives students practice in how to use brackets and what they are for.

For this exercise, use curved brackets ( ), not square brackets [ ].

Discuss the fact that everything in the bracket is a package and must be treated *all* in the same way. Split students into groups to explore this further:

- Give each group cubes of different colours.
- Using the initial letter of the colour of the cube ask them to make this expression: $2r + 3g$

- Ask students how we could show these cubes doubled.
- Show students how this is written mathematically: $2(2r + 3g)$
- Engage students in a discussion about what brackets do.
- Discuss the fact that the 2 on the outside of the bracket means multiply everything in the bracket by 2.
- Ask students then to multiply out their expressions by doubling the cubes.
- Write the equation.
  $2(2r + 3g) = 4r + 6g$

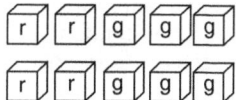

Taking it further

Ask students, 'What would happen if you wished to square your expression? What would happen if you wished to double your expression and add 1? How would you write this out mathematically?'

Contraction and expansion

'From large to little and back again.'

This activity demonstrates the relationship between expanding brackets and factorising. Seeing the process as equal and opposite can help a fuddled brain.

This activity helps students to understand the process of finding common factors in an expression and the process by which the common factors are extracted and put on the outside of the bracket.

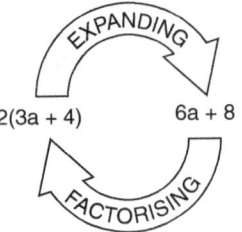

$2(3a + 4)$ $6a + 8$

- Give students in groups the expression $12b + 8g$.
- Give out sets of coloured cubes in green and blue.
- Ask students to model the expression in coloured cubes: b for blue and g for green.

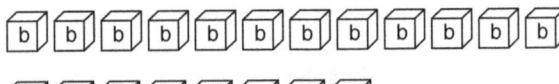

- Ask the students to think of a factor of both 12 blue and 8 green cubes. Students may suggest 2 or 4. See Idea 16 if students struggle with this.
- Ask students to split both the 12 blues and the 8 greens into two equal groups. They should see that they can split them further.

Show them that they have found the highest common factor (HCF) by dividing them into 4 groups.

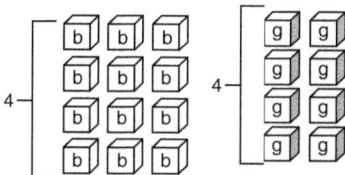

- Ask them to write the expression using a bracket: $4(3b + 2g)$
- Ask students now to multiply the HCF outside the bracket with both terms within the bracket and model the answer with the cubes.

Taking it further

Give students the following expressions to try. Ask them to find the HCF for each expression:

- $6x + 18f$
- $4x + 28g$
- $5ab + 10ac$
  [Can students identify a as a common factor?]
- $8ab + 10bc$
  [Can students identify b as a common factor?]

IDEA 75

Express and equate

'What is the difference?'

Students will be aware that an expression contains variables, coefficients and signs but no equal sign. Equations contain an equal sign (=). It shows the value of the expression on the left side is equivalent to the value on the right-hand side.

Students in this activity will have time to think about the differences between an expression and an equation and be clear about each one.

Give students the Venn diagram below. Ask students to put these phrases in the correct part of the Venn diagram:

- *No equal sign*
- *Has an equal sign*
- *Cannot find value of the variable*
- *Contains numbers*
- *Contains constants*
- *Contains variables*
- *Variable can be found*
- *Contains brackets*

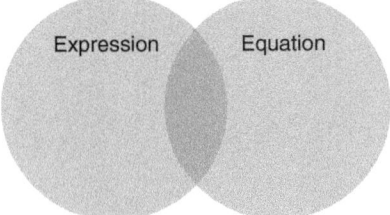

Construct an equation

'Now we get to the exciting bit.'

At last, we begin to construct an equation from a given set of events.

Students should be reminded that an equation contains an equal (=) sign. These activities give real-life examples of constructing equations.

Activity 1

Remind students that they solve equations every day. For example: I am going to the cinema and I have £20.00. The ticket costs £15.00. How much do I have left for popcorn? Explore with students how to write the equation.

- Can students identify what the unknown is? [The cost of the popcorn]
- Can students suggest what they should do to find the cost of the popcorn? [£20.00 take away the ticket cost.]
- Can students write these equations in the two ways above using P for popcorn? [$20 - 15 = P$; $P = 20 - 15$]

Activity 2

Students use a shopping activity to identify the unknown item and find its value. For example: I go to the shop with £20.00 and I buy an item. My change is £16.00. How much did I spend?

- Give students an array of pictures of supermarket items (see online resources).
- For each item students should write an equation, changing the amount of change needed each time. [$£20 - t$ (tea) $= £14$]
- Students should then try to write the equation with the unknown item on its own on the left of the = sign. [tea $= £20 - £16$]

> **Bonus idea** ★
>
> Discuss with students if it matters which way round the equation is, i.e. whether there is a difference between, for example: $20 = 6 + ?$ and $? + 6 = 20$. Many students with numeracy difficulties find it hard to understand that the equations are the same, whichever arrangement is used.

Which operation?

'I couldn't work it out until I used the coins.'

The students practise identifying and using multiplication and division to solve an equation.

The use of real coins can help to illustrate this concept:

- Lay out three 20-pence coins and check students understand that they should multiply 3 by 20 to find the total amount of money.
- Students should now construct an equation: Amount of money = 3 x 20 pence or 3 x 20 pence = 60 pence
- Ask students how to work out the values of the coins if they were not known, for example:

- Ask students to write the equation: $3x = £1.50$. Ask students if they can identify the operation.
- Now ask students to try this example:

- Ask students to write the equation: $6x = 60$ pence. Ask students what the operation is.

Equations again

'Having a visual picture helped me remember.'

Students will now discover what happens when another element, a constant, is added to an equation.

The use of coins here is both visual and concrete and illustrates this type of equation well. Show the students the following example: Three of the same coin plus 7 pence equals 67 pence. What is the value of the coin that is represented by x?

Demonstrate with real coins if possible:

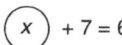 $+ 7 = 67$

- Ask students to write the equation: $3x + 7 = 67$
- Students may see straight away that the coin is worth 20 pence and that 3 of them make 60 pence. If not, they need to solve the equation (see Idea 81).
- Ask them how we can achieve the coins being on their own on the left side of the equation.
- Students should see that they need to remove the 7 pence from both sides in order that the 3 coins equal 60 pence.

Repeat the exercise with 10-pence coins:

 $+ 5$ pence $= 45$ pence

- Ask the students to write the equation: $4x + 5 = 45$
- They will see that the coin is worth 10 pence.
- Ask students to then see if they can work out a rule to solve an equation with an added constant.

Taking it further

Conduct the same activity but now with a negative constant. For example:

x x x $- 7$ pence $= 53$ pence

Ask the students to write the equation:

$3x - 7 = 53$. If students know that three 20-pence coins equal 60 pence then they will see that they need to add 7 pence to both sides.

Algebra is a balancing act

'Scales fall off the eyes, or do they?'

Simple kitchen scales are often used to illustrate how equations work in algebra and they are helpful as they are so visual.

Ask students to try to work out, using the diagram, what *x* would be in this equation. What is the operation to use?

$3x = 21$

In this exercise students will see that the two sides of the equation have to match and hence can begin to see how to manipulate the two sides so that the unknown can be isolated and solved.

- Discuss with students the fact that any mathematical statement involving the = sign must balance.
- The balance is maintained as long as both sides have the same value. See Idea 10.
- Begin with simple equations. This can follow on well from Idea 11.
- Give students pictures of simple kitchen scales on worksheets (see online resources).
- Give students the equation $3x = 60$. Ask them to draw on the scales what they have in the equation, explaining that = is represented by the fact that it balances.
- Can students now solve the equation $3x = 60$? Students should be able to find that x is 20p.
- Now ask students to solve the equation $3x + 7 = 67$, writing each side on the scales.
- Ask the students what happens to the right side of the scales when the 7 is taken from the left side.
- Ask what we need to do to balance the scales once again. They will hopefully conclude that 7 needs to be taken from the right side.

Two sides with an unknown

'What! I can't do this. You are not supposed to have unknowns on both sides!'

Unknowns on both sides of an equation can cause confusion. Students should understand what the unknowns may stand for. They should then try to solve the equation using scales once again.

Ask the students what $5x - 2 = 3x + 4$ could stand for. They may say: 'The cost of 5 bananas minus 2 pence is the same as the cost of 3 bananas plus 4 pence. We need to find the value of x, which is the cost of a single banana.'

Teaching tip

Demonstrating this activity initially with real scales is very useful.

Ask students to write the equation on the scales:

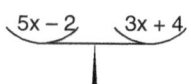

For these equations it is worth deciding on which side you wish to collect the variables (the x's) and which side the constant numbers. Demonstrate that collecting the variables on the right side may involve negatives numbers. Here, it is better to collect the variables on the left side.

Ask students to move $3x$ from the right scale and say what may happen to the scale. Now ask them to remove $3x$ from the left scale and write what they now have. Repeat the exercise for the constants. They should be left with $2x = 6$. Students should work out that $x = 3$. Make sure that they show each step in the calculation.

$$5x - 2 = 3x + 4$$
$$5x - 2 - 3x = 3x + 4 - 3x$$
$$2x - 2 + 2 = 4 + 2$$
$$2x = 6$$
$$\frac{2x}{2} = \frac{6}{2}$$
$$x = 3$$

The sequence of working

'I know the answer and just put it down.'

There are many students who do not write down their working in a structured way when completing algebra, which causes inevitable confusion.

Give students templates for completing the following equations which must be filled in for every step, even if they know the answer. The templates are available in the online resources.

$x + 3 = 10$

$x + 3 - 3 = 10 - 3$ Isolate the x by taking
$x = 7$ the constant 3 from both
                                         sides.

$3x = 60$

$\dfrac{3x}{3} = \dfrac{60}{3}$ Isolate the x by dividing
$x = 20$ both sides of the equation
                                                   by 3.

$3x + 3 = 63$

$3x + 3 - 3 = 63 - 3$ Isolate $3x$ by taking the
$\dfrac{3x}{3} = \dfrac{60}{3}$ constant from both sides.
                                              Isolate the x by dividing
$x = 20$ both sides by 3.

$5x + 2 = 3x + 10$

$5x - 3x + 2 = 3x - 3x + 10$
$2x + 2 - 2 = 10 - 2$
$\dfrac{2x}{2} = \dfrac{8}{2}$
$x = 4$

Isolate the variables on the
left-hand side. Subtract
the constant 2 from both
sides. Isolate x by dividing
both sides of the equation
by 2.

Measurement basics

Part 8

Memorable measures

'Having a physical reference helps students understand what measurements mean.'

Help students to derive personal reference points for units of measurement so that they can envisage what a length or distance looks like. It is an invaluable skill for making judgements in everyday life.

Measurement makes more sense if you have a sense of the scale of the commonly used units such as millimetres (mm), centimetres (cm) and metres (m). Early units of measurement were related to the human body. About 5,000 years ago, the cubit was the distance from the elbow to the tip of the middle finger. From this evolved measurements related to other parts of the body. A foot was the length of the average foot, later standardised as 12 inches; a pace was the length of one step, later formalised as 1 yard, which was the same length as 3 feet.

Students work in pairs to establish portable measuring equipment related to their own bodies. The measurements need to be of practical use for quick estimates.

- Each student writes down an estimate for parts of their body that will measure 1 mm, 1 cm, 10 cm and 1 m.
- Check each measurement using a ruler or a tape measure.
- Discuss how easy it is to use the measurements. This stage is important. There are no wrong answers, merely some that are more practical than others.
- Summarise individual ideas in a table on the board. People are different sizes so there will be variation.
- Discuss why standardised units of measurement are necessary.

How big is that number?

'Sometimes it is impossible to imagine the size.'

A sense of scale helps people understand what very large quantities mean. Cut them down to an intelligible size by considering them in ways that make sense at an individual level.

Students explore the true enormity of million, billion and trillion by relating them to real-life measurements. The teacher gives examples, then students work in pairs to find their own.

Numbers in the news

Check the news for large numbers and then put them into context:

- The richest person in the world earned £26 billion in 2017. Express that as an hourly rate based on a 40-hour working week for 50 weeks. [£13 million per hour]
- 5 trillion plastic bags are produced each year. There are about 7 billion people in the world. If everyone had the same number of plastic bags, how many would they use each day? [2 plastic bags per person per day]

Units of measurement

Ask students to put numbers in a different context by adjusting the units of measurement. Use time as an example: a second seems a very short time until there are lots of them.

- 1 thousand seconds ≈ 15 minutes
- 1 million seconds ≈ 12 days
- 1 billion seconds ≈ 30 years
- 1 trillion seconds ≈ 30,000 years

Students can try the same with distance: a metre is one large step but where would that step take you if there were many? Students put the distances into context to gauge their size. (See teaching tip.)

Teaching tip

Some example distances to discuss with students include:

- 1 thousand metres = 1 km. Establish a distance of 1 km in your neighbourhood using a trundle wheel.
- 1 million metres = 1,000 km. London to Berlin is approximately 1,000 km.
- 1 billion metres = 1,000,000 km. About 25 times around the Earth.
- 1 trillion metres = 1,000,000,000 km. Roughly seven times the distance from Earth to the Sun.

Taking it further

Billions and millions also appear in measurements of minute quantities. The nanometre is used to measure things that are very, very small, such as atoms. A nanometre is one billionth of a metre (0.000000001 m), which is 1×10^{-9} m in scientific notation. Ask students to work out how tall they are in nanometres or find out how thick a human hair is.

In equal measure

'The link between metres, centimetres and millimetres is not obvious to those with numeracy difficulties – even though they appear on the same ruler or tape measure.'

Students separate the measurements shown on a tape measure into three distinct categories by measuring lengths and depicting them on a number line.

Teaching tip

Some tape measures do not show the zero at the beginning; it is implied. Make sure that students understand this.

Consider each unit of measurement – metres, centimetres and millimetres – separately to untangle the conceptual confusion arising from the different units. The tape measure is an efficient way of compressing information into one measuring instrument. Unfortunately, it is visually confusing for those who do not grasp its meaning and the difficulties compound unless the structure is taught explicitly.

Students work in pairs to measure objects and record the information on their number lines. They require a tape measure marked in m, cm and mm. Do not use a metre stick.

Activity 1

- Establish that 1 m = 100 cm = 1,000 mm.
- Ask students to select an object to measure 1 m.
- Students measure the length to establish exactly what 1 m looks like. Discuss with them the importance of the zero at the beginning of the tape measure. This is the starting point for the measurement.
- Students draw a number line. They use short vertical lines to establish the beginning and end of the measurement and mark these points 0 and 1 m.
- Students now measure exactly the same distance. This time they focus on the centimetre units of measurement.

- They draw a second number line exactly the same length as the first, marking the beginning and end points 0 and 100 cm.

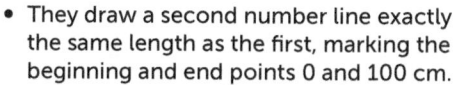

- Then, they measure the distance again, this time focusing on the millimetre.
- Finally, students draw a third number line exactly the same length, marking the beginning and end points 0 and 1,000 mm.
- Discuss with students the fact that the lines represent the same length and what the different units of measurement mean.

Activity 2

In this activity, students learn to convert m to cm to mm.
- Students select an object less than 1 m in length and measure it in metres.
- They mark the length on the number line showing 1 m from Activity 1. Discuss with students why the length will be written as a decimal fraction of 1 m.
- Now, students repeat the process for measurements in cm and mm.

Activity 3

This activity brings it all together.
- Students measure another object that is between half a metre and 1 m long.
- They draw a number line marked 1 m and locate the new measurement on it.
- They write the value as a decimal fraction of 1 m below the number line as shown.
- Students then measure the same distance in cm and position it on the same number line. They write the value above the number line.
- Students then repeat the process for the measurement in mm.

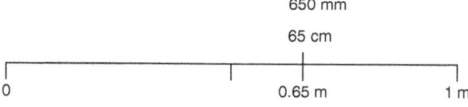

> ### Bonus idea ★
>
> Develop students' ability to visualise relative measurements by allowing them plenty of time to play around with lengths and areas of objects. Record them in different units on number lines and area diagrams. Interpreting visual data is a vital skill in reading graphs. Students need to be able to spot intentionally distorted measurements on graphs (both linear and pictorial). Distorted graphs misuse statistics to present a particular viewpoint.

Average

'Average does not mean ordinary. It summarises a group of numbers.'

The real world requires easy ways to summarise data. Averages are an important way of doing that. As well as the average you need to know the range.

An average is one number that is used to represent a larger collection of numbers (known as data) to make it easier to understand. The data is gathered from measuring things. There are three kinds of averages – the mean, the median and the mode. Encourage students to discuss the pros and cons of each type of average.

First, students should learn how to calculate each kind of average, then they can look behind the numbers to find out what averages tell us and what they leave out. Complete the following activity as a class:

Find the range and the three types of averages in this data set: 15, 2, 12, 3, 8, 8, 9, 2, 3, 2, 13. Record the results in a table.

The **range** is the distance between the lowest and highest number in the data set. To find it:

- Subtract the lowest number in the data set from the highest number.

The **mean** is the average that most people think of when they use the word average. To find it:

- Count how many terms there are in the data set.
- Add all the numbers.
- Divide the result by the total number of terms in the group.
- The result is the mean value for the group.

The **median** is the middle number in a set of numbers. To find it:

- Arrange all the numbers in order of size, starting with the smallest.
- Count the number of terms and find the midpoint of all the terms.
- If there are an even number of terms, the midpoint will be between the two numbers in the centre of the group.

The **mode** is the number that appears most often. To find it:

- Arrange the numbers in order of size, starting with the smallest.
- Underline the numbers that are the same size.
- The mode is the number that appears most often.

Data set	15, 2, 12, 3, 8, 8, 9, 2, 3, 2, 13	
The range	15 − 2 = 13	Range = 13
Mean	Number of terms: 10 Total value of terms: 15 + 2 + 3 + 3 + 8 + 12 + 9 + 2 + 8 + 2 + 13 = 77 77 ÷ 11 = 7	Mean = 7
Median	Arrange terms in order of size: 2, 2, 2, 3, 3, <u>8</u>, 8, 9, 12, 13, 15 There are 11 terms. The middle number is 8.	Median = 8
Mode	<u>2, 2, 2,</u> 3, 3, 8, 8, 9, 12, 13, 15 The number that appears most often is 2.	Mode = 2

Weighty matters

'It has been helpful to experience the feeling of different weights.'

In this activity, students build a memory bank of the size of units of mass to provide a benchmark against which to check quantities in calculations. It is invaluable for getting a quick sense of whether an answer is sensible or not.

Students guess the mass of various objects, then weigh them and discuss the results. Then, students compile a list that they find meaningful to envisage different weights. Students work in small groups so that each student has a chance to discuss their ideas.

Guess the mass

- Provide five objects of varied sizes, e.g. pen, calculator, textbook.
- Each student guesses the weight of the objects and records their estimates in a table.
- They weigh each object and record the mass.
- Discuss the variation between the estimates and the actual measurement.

My quick mass monitor

Each student compiles a list of five everyday objects to gain a rough idea of what mass represents from small (1 g) to large (1 tonne). They include their own weight as a benchmark. It is impractical to measure large masses so they can use published information to establish some amounts.

Example:

Student's weight	50 kg
Small apple	100 g
Large bag of sugar	1 kg
Bicycle	12 kg
King size mattress	100 kg
Compact car	1.5 tonnes

Graph stories

'Join the dots for a picture of the story behind the numbers.'

What is a line graph for? It is a diagram that shows the relationship between variable pieces of information as a line. The line results from joining data points corresponding to the x and y coordinates. A graph makes it easy to take in data at a glance.

Students with spatial difficulties find graphs very difficult to interpret and to construct. They have problems interpreting the values that the intervals on the axes represent. They find it particularly hard to work with the vertical axis. Ensure that students can work confidently with number lines before starting graph work.

Interpret a graph

Consider data in a table and then on a graph to demonstrate the power of a picture. You can use the table and graph showing the price of gold from 2002 to 2015 (see online resources). The graph shows the dramatic rise in the gold price during the global financial crisis.

- Project a table of figures on the whiteboard for 15 seconds.
- Switch it off and ask students to comment on any changes they saw in the numbers.
- Project a picture of a graph of the data for 15 seconds.
- Ask students if they noticed any changes in the graph.

Draw a graph

Students each draw their own graph to show monthly rainfall. Discuss the importance of using a sensible scale that fits the space available so that it is easy to read. Provide students with the data table in the online resources. An example graph is also available.

Time and motion

'Time moving is a hard concept to grasp.'

Time measurements are based on natural rhythms that are repeated constantly. Pose questions to help students link the familiar components of time to the movement of the Sun and the Moon.

Taking it further

In 1967 the second was chosen as the SI (standard international) unit of time. The length of one second is based on how many times a particular type of atom vibrates.

Students often find time a difficult topic. Misunderstandings arise because the measurement of time involves working with several different number bases (though this is seldom explicitly taught) and time is displayed in a variety of different formats. This idea and Ideas 89 and 90 are designed to reduce anxiety about time by looking at the origins of time measurement.

Units of time

Students can work as a class or in small groups. Allow students five minutes to write down all the units of time they can think of and then write down what each unit of measurement is based on. Questions to help:

- What is the unit?
- What natural pattern is it based on?
- How long is it?

Bonus idea ★

Explore the history of time for a fascinating project involving geography, physics and history.

Students summarise their ideas in a table:

Unit	Based on	Length
Year	The length of time the Earth takes to orbit the Sun.	365 days
Day	The time taken for the Earth to rotate once on its axis.	1 day = 24 hours
Hour	The day is divided into 24 hours.	1 hour = 60 minutes
Minute	The hour is divided into 60 seconds.	1 minute = 60 seconds
Month (lunar)	The length of time the Moon takes to orbit the Earth.	28 days (lunar month)
Month (calendar)	The year is divided into 12 months. Most months are slightly longer than the lunar month in order to use all 365 days.	Months vary between 28 and 31 days.
Week	A collection of days that is one quarter of the Moon's orbit.	7 days
Fortnight	Two weeks.	14 days

The language of clocks

'The language of time is stuck in the world of analogue timepieces, although most people now use digital displays.'

Difficulties with doing time calculations are often rooted in a failure to understand the basics of time. Reading and understanding time often presents a problem because the language used to describe time derives from the position of the hands on the clock face. Another complication is that the clock face has two different scales – 12 intervals for hours and 60 intervals for minutes.

This activity links the clock face to the number line and demonstrates why 'half', 'quarter', 'to' and 'past' are used. Students then relate the length of time to fractional proportions on the number line. Often, confusion about telling the time starts with a half. Students who have learned that 0.5 is half of 1 whole and have used the 'pizza' model to show 0.5 on a circle, do not understand why half an hour is labelled as 0.30. They do not realise that the dot in 0.5 is a decimal point, whereas the dot in 0.30 is a separator between minutes and hours (see Idea 91).

Line time

- Give each student two strips of paper of the same length and about 2 cm wide.
- Ask them to draw a number line on Strip A and label the ends 0 and 1.
- They now divide the strip into quarters and mark the intervals on the number line as fractions.
- They draw a number line on Strip B and label the ends 0 and 60.
- They fold Strip B into quarters and mark the appropriate values at the intervals on the number line (15, 30 and 45).
- They compare Strip A and Strip B.

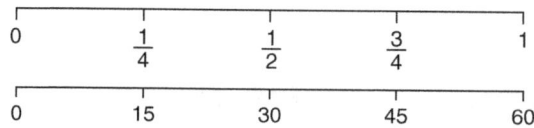

Taking it further

Problems arise because two different measures of time are shown on the same clock face – hours and minutes. Ask students, 'Why do the two scales work together?' The answer is that, although they have different bases, the scales work together because 12 is a factor of 60.

Circle time

- Now students join the ends of each paper strip to form two circles. They hold each circle so that the join is at the top.
- They draw a diagram of each circle and label it with the values from the number line. Note that zero can be omitted as that point is both the starting and the end point.

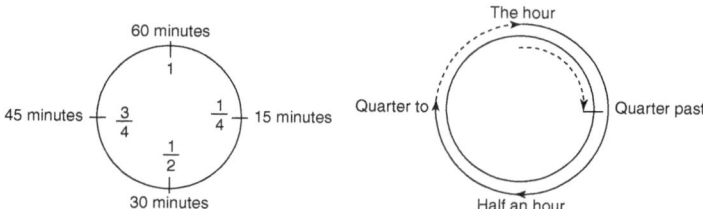

The language of quarters and half

- Students draw a number line to show the relationship between the fractions, the minutes and the expressions.

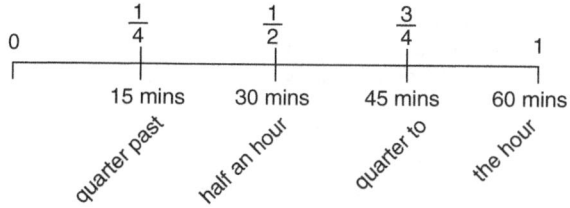

They draw up a table to show that:

- 60 minutes = 1 hour
- 15 mins = a quarter of an hour
- 30 mins = half of an hour
- 45 mins = three quarters of an hour

Digital and analogue time

'Time on the clock face does not link intuitively to digital displays.'

Failure to understand the basics of time leads to difficulties doing time calculations. Students need to read analogue time on a clock (see Idea 89) and then they need to be able to relate analogue and digital time and write both forms. Tabular and linear diagrams make the link clear.

Students use a number line to reconcile analogue time and digital time, then record information in a table. The linear representation also makes it easier to do time calculations since most time problems involve either finding out how much time elapses, or when a time interval starts or finishes. The number line provides a structured way to display information and do the calculation. It is also a good model for the concept of time as a continuous linear measurement.

Analogue and digital

- Discuss what is the same and what is different about the measurement of time using an analogue clock and a digital clock.
- Questions to prompt discussion:
  - What does the word 'day' mean?
  - How long is a day?
  - When does the time measurement of a day start?
  - How long is one complete cycle on the clock?
  - How do you differentiate between the first and second 12-hour period of the day?
  - How long is one complete time cycle on a digital clock?

Key points to consider: the word 'day' means a 24-hour period of time from midnight to midnight. It also means the hours of daylight as opposed to night-time. The analogue clock records a day as two 12-hour time periods. A digital clock can show the full 24-hour period of time or it can record two 12-hour periods. The 12-hour cycles are designated as morning and evening and denoted by the abbreviations 'am' and 'pm'.

Time on the number line and in a table

In this exercise, students write the 12-hour times below the line and the 24-hour times above the line. Note that 12 o'clock does not specify noon or midnight so students will need to write both times. However, do not instruct them to do this; wait until after they have done the exercise. Each student draws their own timeline following these instructions:

- Draw a line marked 0 at one end and 24 at the other.
- Mark the following times on an empty number line:
  ○ 12 o'clock
  ○ 3 o'clock in the afternoon
  ○ 5 o'clock in the morning
  ○ the time school starts
  ○ the time you went to bed yesterday.

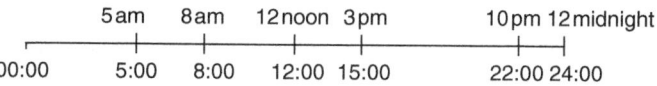

- Write the times above in order from earliest to latest.
- Record them in a table showing the 12-hour clock and 24-hour clock.

12-hour clock	24-hour clock
5 am	05:00
8 am	08:00
12 noon	12:00
3 pm	15:00
10 pm	22:00
12 midnight	24:00

Bonus idea ★

Use clocks to entice students into the world of electronics, which affects almost every aspect of our lives. There are two broad categories in electronics – analogue and digital – and clocks neatly summarise the difference. Do a project on how electronics has affected the world of sport where times can be measured in hundredths of a second.

The calculator trap

'Beware the time trap that is caused by the dots between the numbers in calculator calculations.'

Clarify the meaning of the separator dots that are used in writing time. They mark the boundary between the hour and the minutes.

Taking it further

Students can refresh their time management skills while they practise recording and converting between analogue, digital and decimal notation. They plan what they want to study, and for how long, during one week (see Idea 3). They record the information in a table. At the end of each day, they use a timeline to show what they achieved and when.

This activity practises writing time in various ways. Some students find the different ways that dots are used in writing time very confusing. The separator between hours and minutes may be a single dot or a colon. These markers are not to be confused with decimal points (see Idea 30). However, the results of calculations on a calculator will show time with a decimal point.

Pose the question: Sam travels 2 ½ hours by bus and 1 ¼ hours by train. How long is the total journey? If the bus leaves at 8.00 am, what time does Sam arrive at his destination?

- Ask students to show the journey times in a table using both clock time and decimal notation.
- They then add the times and explain why 3.75 hours is the same as 3 hours 45 minutes.
- They use a table and a timeline to show the total journey time and the arrival time at the destination.

	Journey time	Clock time	Decimal notation
Bus	2 ½ hours	2 hours 30 minutes	2.5 hours
Train	1 ¼ hours	1 hour 15 minutes	1.25 hours
Total time taken	3 ¾ hours	3 hours 45 minutes	3.75 hours

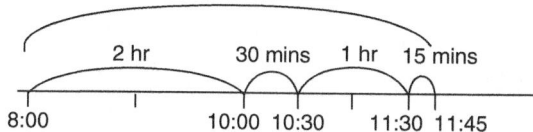

Journey time answer: The journey time is 3 hours and 45 minutes. 3.75 hours means 3 hours and $\frac{75}{100}$ of 1 hour.

Since 1 hour = 60 minutes, find 0.75 of 60 minutes:

0.75 x 60 = 45 minutes.

So 3.75 hours is the same as 3 hours 45 minutes.

Arrival time answer: The arrival time is 11.45 pm.

Talking time

'The vocabulary associated with telling the time frequently causes problems.'

Familiarise students with the equivalent positions of numbers and words used in measuring time on a clock face. Get students talking about time and the different ways that the same time can be expressed by putting out cards in the relevant positions on a clock face.

The fact that hours and minutes are represented on the same clock face confuses some students. This activity, for an individual student, makes it clear that the same point on the clock face can represent both hours and minutes and be described in several different ways, e.g. the position marked 3 may denote the hour 3 or 15 minutes. These are the values at that point. It also indicates the time relative to an hour such as 'quarter past'.

Print a pack of 52 cards that have various descriptions of time on them, plus four *Time is up!* cards (see online resources). The aim of the game is to put all the cards in the correct position in the circle before all the *Time is up!* cards are exposed. To play:

- Shuffle the pack then arrange 12 cards, face down, evenly in a circle and put one card in the centre of the circle.
- Deal the rest of the cards face down, so that there will be four cards on each pile.
- Turn up one card from the middle. Place it face up in the correct position on the clock face, next to the relevant pile of cards.
- Turn up the top card from the adjacent pile and put this card in the correct 'clock' position.

- When a *Time is up!* card is exposed, exchange it for one of the cards in the central pile.
- Continue play until all the cards are in the correct places.
- Beware! Once all four *Time is up!* cards are in the centre, the game ends and starts again.

Start position

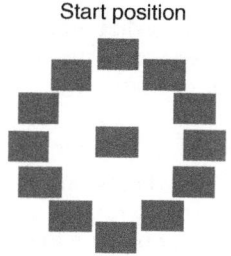

A game in play

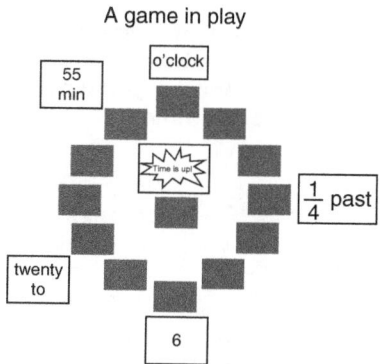

Bonus idea ★

Play the same game with months of the year cards. Pupils with numeracy difficulties may not know the number relating to each month of the year. Play the same game with numbers 1 – 12 and the months of the year. (There is a template for the cards in the online resources.)

Distance-time graphs

'Changes in the slope of the line make it easy to see changes in the speed.'

Speed is a measure of how fast something moves. In a distance-time graph, the gradient of the line indicates the speed of the object. The greater the gradient (and the steeper the line) the faster the object is moving.

Taking it further

Give students a graph in which the gradient of the slope changes. What does this mean and how should they work out the speed? Ask students to make up a running story involving changes in pace and plot it on a graph.

In this activity students discuss what the graph means and work out the speed of a runner. Then, they construct a graph to compare the speed of two runners. (This builds on the work on ratio; see Idea 51.)

First, show the class a distance-time graph:

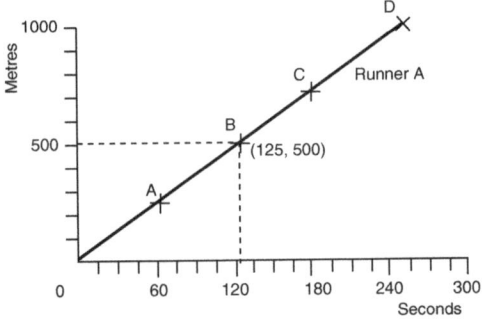

Discuss the composition of the graph. Ask students to identify the axes representing distance and time. Discuss the meaning of the word 'gradient' and the way in which speed is expressed as distance travelled in a given time unit.

On their own A4-size copy (see online resources), students write 'Distance' and 'Time' on the appropriate axes. They find the value of the x and y coordinates at each marked point.

Students suggest, in their own words, how you calculate the speed for Runner A. Give prompts:

- What does speed mean? [Speed is distance travelled in a specified time.]
- How do you express the units of speed in this example? [m per sec]
- Can you work out a formula for speed? [Speed = Distance divided by Time]
- Now make the calculation:

$$S = \frac{D}{T}$$

$$S = \frac{500}{125} \text{ m per sec}$$

$$S = 4 \text{ m per sec}$$

Comparing graphs

Ask students to plot the speed for Runner B on the same graph using the following coordinates for (x,y): (33,250), (99,750) and 132,1000).

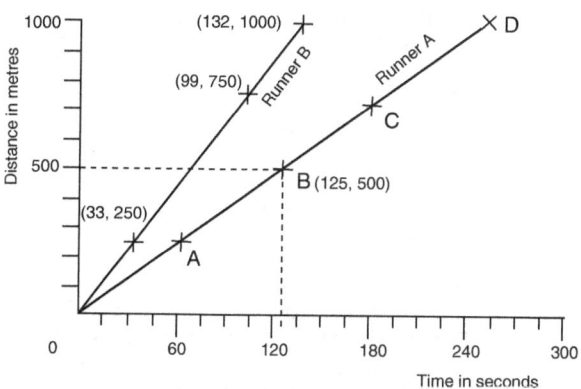

Explain what the gradient of the new graph tells you about Runner B's speed in relation to Runner A. Now ask them to calculate the speed for Runner B. [7.58 m per sec to 2 d.p.: the world record set in 1999 by Noah Ngeny.]

Bonus idea ★

The Cartesian grid (or graph) was devised in the sixteenth century when many new mathematical ideas were developed. Explore the history of these changes, which included accurate mechanical clocks, artillery weapons and maps for world exploration.

Temperature

'A great way to show that negative numbers are real.'

Students may have difficulty doing calculations involving temperature change because they do not understand the scale on a thermometer. A further problem arises because there are two different scales for measuring temperature – Centigrade and Fahrenheit. Students need to understand the Centigrade scale before using Fahrenheit.

Teaching tip

Make sure that students can locate positions on number lines and use number lines for calculations before embarking on temperature measurements.

Temperature is a measure of how hot or cold something is. The temperature is recorded on a thermometer with a scale that is calibrated in degrees, which may be negative or positive. Key points on the Centigrade scale are 0°C (zero) and 100°C.

A sense of temperature

Put a picture of a Centigrade scale on the board. Although thermometers are often shown in a vertical orientation, it is helpful to show it as horizontal as that relates directly to the number line.

- Give each student a blank number line. Ask them to mark -100°C, 0°C and 100°C.
- Students locate and mark the following temperatures (give them in a random order): 4°C, -18°C, 57°C, -89°C, 37°C,
- Now ask students to match the following descriptions to the appropriate temperatures:
  - Water freezes
  - Human body temperature
  - Lowest temperature in the world
  - Inside a domestic freezer
  - Inside a fridge
  - Highest temperature in the world
  - Water boils

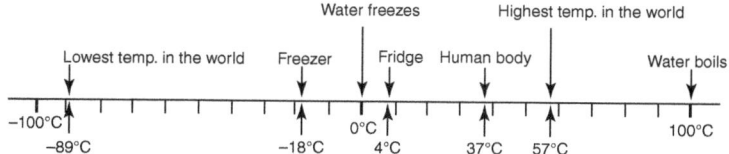

Lowest temp. in the world — Freezer — Water freezes — Fridge — Human body — Highest temp. in the world — Water boils

−100°C   −89°C   −18°C   0°C   4°C   37°C   57°C   100°C

Calculate temperature change

Which city has the largest temperature change?

- Check the weather forecast for the temperature variation in several cities around the world on the same day.
- Record the data in a table.
- Calculate the temperature variation for each city.

City	Highest temp	Lowest temp	Temperature change
Johannesburg	26 °C	17 °C	26 °C − 17 °C = 9 °C
London	10 °C	3 °C	10 °C − 3 °C = 7 °C
Moscow	−1 °C	−8 °C	−1 °C − (−8 °C) = 7 °C
Mumbai	31 °C	23 °C	31 °C − 23 °C = 8 °C
New York	8 °C	−2 °C	8 °C − (−2 °C) = 10 °C

- Students sketch each calculation on a number line.

Example: New York

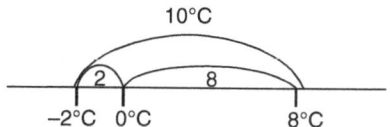

10°C

2   8

−2°C   0°C   8°C

> **Bonus idea** ★
>
> Do a geography project on average annual temperature changes and show them on a graph. The NASA Scientific Visualization Studio has an interactive map showing global temperature changes from 1884 to 2018 at https://climate.nasa.gov/vital-signs/global-temperature.

Interesting
investigations

Part 9

Squares and odd numbers

'Deriving the formula from the model makes it much easier to understand.'

The ancient Greeks were fascinated by figurate numbers, which can be modelled as regular geometric patterns. A number multiplied by itself is a square, for example, hence the term square numbers.

Square numbers

Students work in pairs to investigate the sequence of numbers: 1, 2, 3, 4 and 5. Provide the following instructions:

- Multiply each number by itself and use counters to model the calculation as an array.

$1 \times 1 \quad 2 \times 2 \quad 3 \times 3 \quad 4 \times 4 \quad 5 \times 5$

- Investigate and discuss what needs to be added to the 5th term to find 6 squared.
- Draw up a table to show the relationship between each number (designated n) and its square (n^2).
- Calculate the difference between each square number and add it to the diagram.

Term	1	2	3	4	5	6
Value of term	1	4	9	16	25	a_6
		+3	+5	+7	+9	

The sum of odd numbers

Now try this activity with students:

- Start with 1 and add consecutive odd numbers to generate a sequence of numbers. Model the results using counters.

1   1 + 3   1 + 3 + 5   1 + 3 + 5 + 7
•   88   888   8888

- Record the results in a table showing each calculation and the value of the term in each position.

Term	1	2	3	4
Value of each term	1	1 + 3 = 4	1 + 3 + 5 = 9	1 + 3 + 5 + 7 = 16

- What do students notice?
- Find the rule that governs the sequence.
- Draw a diagram that emphasises the relationship between the square and the odd number that is required to form the next square number.

Taking it further

Students discuss the relationships and experiment to find a formula to generate any number in the sequence. The formula is $a_n = (n - 1)^2 + 2(n - 1) + 1$

159

Number triangles

'This has helped me be fascinated, not fearful.'

A triangular number can be modelled as a grid of objects in which each row is one more than the row above. The sequence generating this arrangement is formed by adding consecutive numbers. The mathematics of triangular numbers has important applications in designing computing networks.

Modelling triangular numbers

- Students use counters to model the first four triangular numbers. They start with one counter, then put counters out in rows, one under the other, to form a sequence of equilateral triangles.
- They draw up a table with a diagram of each number and the triangle.
- They write the equation that makes it clear how each number is generated.

Triangle number	1	2	3	4
Model of triangle	•	••	•••	••••
Equation	$1 = 1$	$1 + 2 = 3$	$1 + 2 + 3 = 6$	$1 + 2 + 3 + 4 = 10$

A rule for any term

Ask students to consider the following questions:

- How many counters will there be in the 5th triangle?
- Is there a general rule to find any number in the sequence?
- Hint: use the area of a rectangle and halve it.

Allow students to experiment before discussing the following model:

- Rearrange the dots to form a right-angled triangle.
- Repeat the pattern and join the two models together to form a rectangle.
- Draw a diagram and label the sides of the rectangle.
- Calculate the number of dots in the rectangle. (Multiply the length by the width.)
- The number of dots in the triangle is half those in the rectangle.
- Write the formula for any triangular number:
  $$\frac{n(n + 1)}{2}$$

Triangle number (n)	1	2	3	4
Model	$1\{$ •o $1 + 1 = 2$	$2\{$ •○○ •○ $2 + 1 = 3$	$3\{$ •○○○ ••○○ •••○ $3 + 1 = 4$	$4\{$ •○○○○ ••○○○ •••○○ ••••○ $4 + 1 = 5$
Dots in rectangle n(n+1)	$1 \times (1 + 1) = 2$	$2 \times (2 + 1) = 6$	$3 \times (3 + 1) = 12$	$4 \times (4 + 1) = 20$
Dots in triangle n(n+1)/2	$1 \times (1 + 1)/2 = 1$	$2 \times (2 + 1)/2 = 3$	$3 \times (3 + 1)/2 = 6$	$4 \times (4 + 1)/2 = 10$

Taking it further

Can students find a quick way to add all the numbers from 1 to 100? The task was made famous by a young child called Carl Frederick Gauss who became one of the most influential mathematicians of all time. Young Gauss realised that there are 50 pairs of numbers that add up to 101. (100 + 1, 99 + 2, etc.). So the sum of the series is 5,050 which is the hundredth triangular number. Can students write a formula and find the sum of all numbers from 1 to 1 million?

From one to eternity

'It felt like going into a tunnel that keeps getting smaller but has no end.'

Students practise adding fractions and showing them on a diagram of a square that represents one whole. This innocent-looking activity has hidden depths. It drives home the fact that the bigger the denominator in a fraction, the smaller the quantity. The series moves tantalisingly close to 1 but it will never reach it.

Taking it further

This series shows the power of geometric progression. Here, halving rapidly reduces the size. Compare it to the effect of doubling in Idea 100.

This simple activity gives students practice in adding fractions but leads to a startling result. Complete this exercise as a class but ensure students do their own calculations and diagrams.

Calculate the series

A sequence is a list of numbers placed in a defined order. A series is the sum of the numbers in that sequence.

- Consider the sequence: $\frac{1}{2}$ $\frac{1}{4}$ $\frac{1}{8}$ $\frac{1}{16}$
- Add the fractions, showing your working then show the calculation in a diagram.
- Draw a square to represent the whole. Show each fraction as a part of the square. Colour each fraction for clarity.

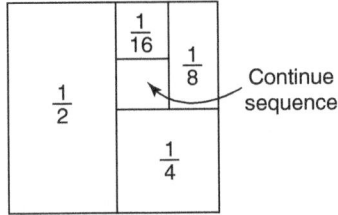

Continue sequence

Continue the sequence

Students work in pairs to find the rule governing the sequence.

- Uncover the relationship between the terms in the sequence. [Multiply a term by $\frac{1}{2}$ to find the next term]
- State the rule for finding the value of the nth term in the sequence. The nth term = $\frac{1}{2^n}$
- Produce a table to depict the rule and show the next two terms in the sequence.
- Continue the series and update the diagram.

Taking it further

Draw a long number line and label the ends 0 and 1. Mark each term in the sequence starting with $\frac{1}{2}$. This exercise highlights the difficulty of measuring very small quantities.

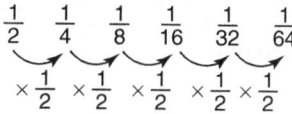

A convergent series

The diagram makes it clear that the series is converging towards 1. However, it will never reach it. Express the fractions as decimals for another way of looking at this fascinating series:

0.5 + 0.25 + 0.125 + 0.0625 = 0.9375

Into another dimension

'I see an icon of three twisted Mobius strips several times a day. It is the recycling symbol.'

Astound students with this simple experiment as a rectangular piece of paper with two surfaces proves to have only one surface. The equipment consists of plenty of strips of paper roughly 5 cm wide and 20 cm long, glue, scissors and a pencil.

Taking it further

Students can investigate what happens when the strip of paper is twisted twice and then what happens if it is twisted three times. Most of us see the icon of the three-twist Mobius strip several times a day.

Bonus idea ★

The Mobius band is a good way to introduce the subject of topology, which describes mathematical spaces. It is important in many fields including understanding DNA in biology, analysing data sets in science and economics, as well as for studying the structure of the universe.

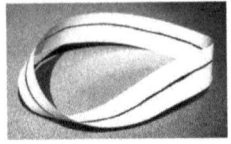

Explain to students that we should start each of the following experiments by answering the question: 'What do I expect to find out?' We should end each experiment by discussing what did happen and why. We want to find out the following:

- How many surfaces are there?
- How many edges are there?
- How many shapes are there?

Task 1

Take one strip of paper and join the ends to form a ring. Draw a line along the middle of the inside of the ring.

Task 2

Take a strip of paper, give it a half-twist and join the ends to form a band. Draw a line along the middle of the inside of the band. (Students will discover the band, known as a Mobius band, has only one surface and one edge. They can deduce this from the fact that the line they draw continues and returns to the starting point.)

Task 3

Cut the Mobius band along the mid-line.

Task 4

Draw a line one third of the distance from the edge of the Mobius band. Cut along the line.

Investigating pi

'What an irrational number pi is. You can't calculate the exact value but it is really useful!'

A circle is a shape where every point on its circumference is the same distance from the centre. There is a relationship between the circumference and the diameter of a circle, which is a constant number called pi, written as π.

In this activity students work in small groups to investigate pi and find its value. They measure objects to provide memorable images of the concept of pi and then calculate its value.

- Students work in pairs. Provide each pair with a flexible tape measure.
- Ask the students to think of circular items available in the classroom such as stationery holders. Alternatively, to save time, supply a variety of circular objects of different sizes: canned fruit, plates, bowls or bicycle tires.
- Students draw up a table with the headings: Item, Circumference, Diameter, Ratio, Pi.
- They measure the circumference (C) and diameter (D) of each object and record the ratio.
- They write the formula for finding pi. [$\pi = \frac{C}{D}$] and divide the circumference by the diameter for each object to find the value of pi to 2 d.p.
- Students discuss their findings. The results should be close to 3.14. If not, students need to find out why.

Taking it further

How does pi help find the circumference of a circle? If pi is found by dividing the circumference by the diameter ($\pi = \frac{C}{D}$) then you can find the length of the circumference by multiplying pi by the diameter. Write the formula $C = \pi D$. Since the radius is half the diameter, it can also be written as $C = 2\pi r$.

Bonus idea ★

Do a history project on pi to discover the importance of pi since ancient times.

Item	Circumference	Diameter	Ratio	Pi (2 d.p.)
Plate	73 cm	23 cm	73:23	73/23 = 3.17
Bicycle tire	175 cm	55 cm	175:55	175/55 =3.18

What do the results show? Try to encourage them to see that, whatever the circumference or diameter, the relationship is the same.

Growth drama

'Would you rather have £1,000 or the sum of one pence doubled every day for a month?'

In this investigation, students work out which is the best rate of pay in the following scenario. A job lasts for 32 days. Employee A is paid £1 on the first day. Each day the amount paid is £1 more than the day before. Employee B is paid 1 pence on the first day. Each day they are paid double the amount from the day before.

Teaching tip

While students start by counting grains of rice, it soon becomes clear that it is more efficient to use other means of assessing the quantity. This is a valuable lesson in understanding that there is often more than one way to measure something.

Model the pay rates on two large chessboards as well as drawing up a table to show the daily amount paid to each employee. Each student also records the information on their own diagram of a chessboard. This whole-class activity requires students to collaborate and organise themselves into small groups to carry out various parts of the investigation. While some students build the models, others work out the accurate amount for each square.

It is impractical to use money to model the amounts, so use grains of rice instead. One grain of rice represents 1 pence. As the numbers get larger, it is easier to weigh a specified number of grains and then to use weight as a way to find the approximate number required. Later, it is more convenient to find out how many grains fill a large box. A box can represent the number of grains so it is not necessary to fill the boxes with rice.

The equipment consists of about 10 kg of rice in packets, 32 black and 32 white bin liners to make chessboards, weighing scales and cardboard boxes.

- Make two chessboards from the bin lines. Cut each bin liner in half to form large squares and join them together.
- Divide the class into two groups: Group A investigates the pay rate for Employee A.

Group B investigates the pay rate for Employee B.

- Each student draws up a table to record the daily rate and two 8 x 8 diagrams consisting of 2 cm² squares (see online resources for a template).
- Students place grains of rice on the squares on the chessboards to model each day's pay.

Group A:

- Employee A earns £1 on day one. Students work out the equivalent amount in grains of rice and count these out to put on the first square. £1 is 100 pence so 100 grains of rice are required. Count out 100 grains of rice. As the numbers increase, it is easier to use weight or volume to assess the quantity.
- Students place the 100 grains on the first square of the chessboard and record the number of grains of rice in the table.
- They continue the model by adding 100 grains of rice to the previous amount to find the new amount.
- When they have worked out how many grains of rice are on the 32nd square, students add up the total number of grains earned and convert it into pounds. Remember 100 grains of rice represent £1.

Taking it further

Work out how many grains are on each square for the whole chessboard. Round numbers to 2 significant figures and write them in standard form. Then, calculate the total number of grains of rice on the chessboard.

Employee A: the first 8 days

Day	Grains of rice on each day
1	100
2	200
3	300
4	400
5	500
6	600
7	700
8	800
Continue to day 32	

Group B:

- Employee B earns 1 pence on day one so students place 1 grain on the first square of their chessboard.
- They record the number of grains in the table.
- They continue the model by doubling the amount on the square to find the next value.
- When they have worked out how many grains of rice are on the 32nd square, students add up the total number of grains earned and convert it into pounds. 100 grains of rice represent £1.

Employee B: the first 8 days

Day	Grains of rice on each day
1	1
2	2
3	4
4	8
5	16
6	32
7	64
8	128
Continue to day 32	